ELEPHANT *Speak*

A Devoted Keeper's Life Among the Herd

ELEPHANT *Speak*

A Devoted Keeper's Life Among the Herd

MELISSA CRANDALL

"The love between two species is not like the love within one."
—Sallie Tisdale

Elephant Speak: A Devoted Keeper's Life Among the Herd

ISBN13: 978-1-947845-10-7

Ooligan Press
Portland State University
Post Office Box 751, Portland, Oregon 97207
503.725.9748
ooligan@ooliganpress.pdx.edu
www.ooliganpress.pdx.edu

Library of Congress Cataloging-in-Publication Data:
Names: Crandall, Melissa, 1957– author.
Title: Elephant speak : a devoted keeper's life among the herd / Melissa Crandall.
Description: First edition. | Portland : Ooligan Press, [2020] | Includes bibliographical references. |
Identifiers: LCCN 2019039261 (print) | LCCN 2019039262 (ebook) | ISBN 9781947845107 (paperback) | ISBN 9781947845114 (ebook)
Subjects: LCSH: Henneous, Roger. | Zoo keepers--Oregon--Biography. | Oregon Zoo (Portland, Or.)

Cover design by Laura Mills and Jenny Kimura
Back cover photo provided by *The Oregonian* (1998)
Interior design by Denise Morales Soto

Printed in the United States of America

For Belle

Contents

Author's Note

Roger Henneous and I met in the spring of 1997, when I was a volunteer at the Washington Park Zoo (now the Oregon Zoo). His devotion to the elephants in his care, and their obvious love for him, affected me so profoundly that twenty years later I searched him out and asked to write his life's story. Had I approached him any earlier with the request, he says he would have refused, but time had mellowed Roger and the opportunity to relive those days beckoned. Our weekly conversations took place from 2015 to 2018. This book contains his memories and reflections, supplemented by explanatory research.

Because the zoo in Portland, Oregon has gone by a variety of names since its inception in 1888, I've used the current title "Oregon Zoo" throughout the book in order to minimize confusion.

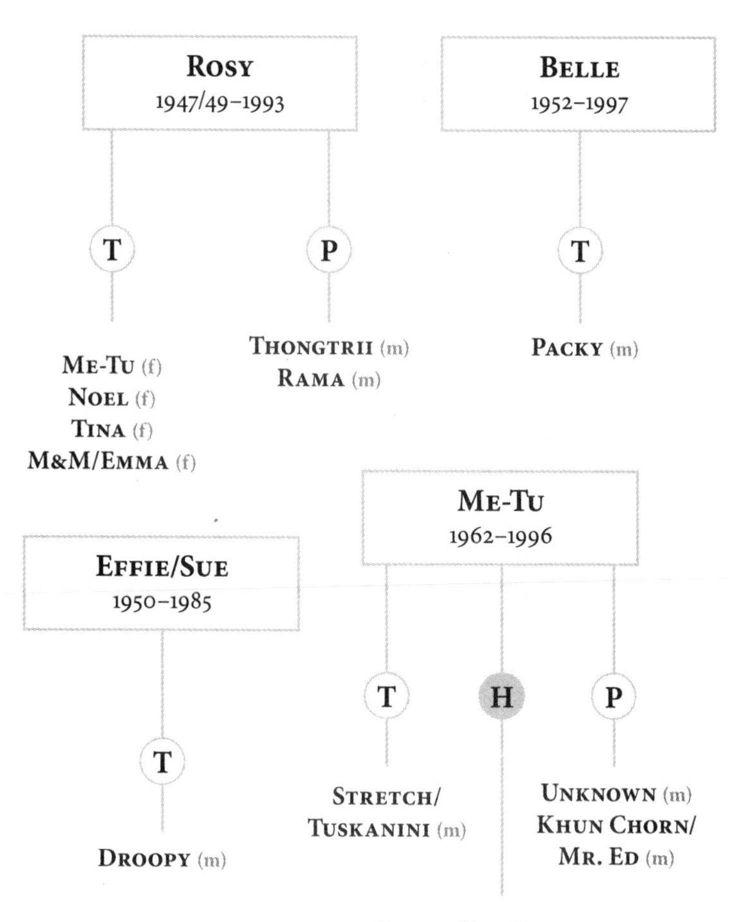

ELEPHANT FAMILY

Tree

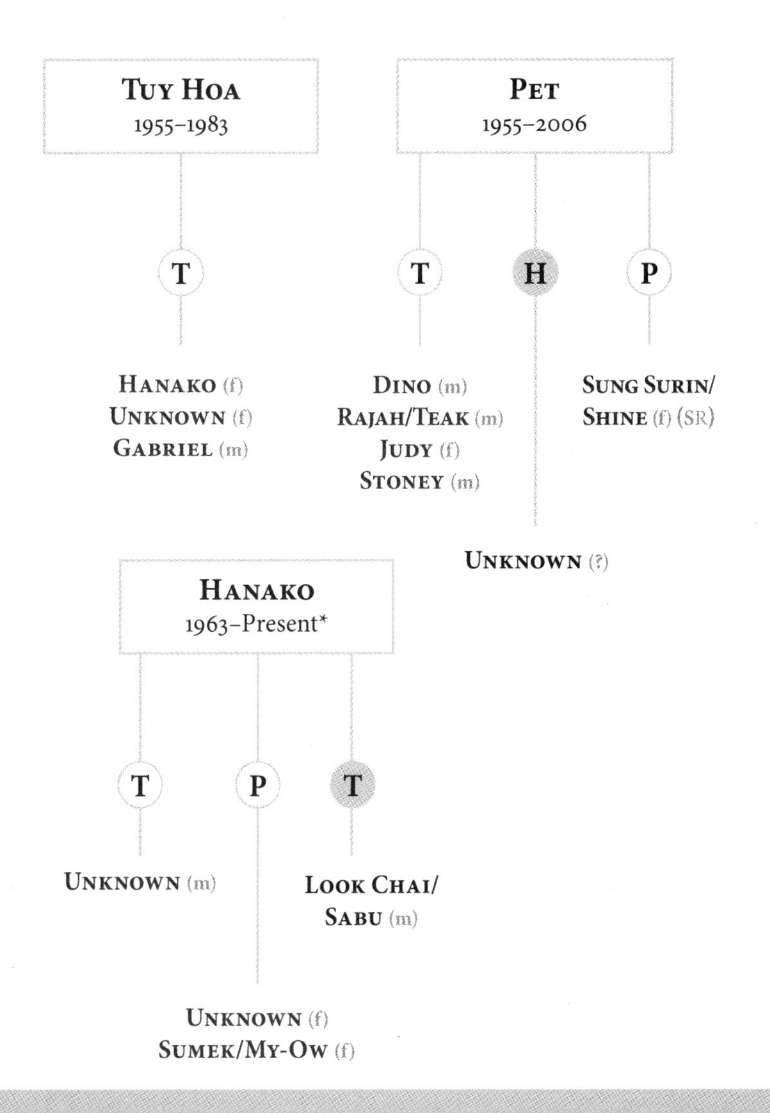

Legend

Present* = 2019

SR = Still resides at Oregon Zoo as of 2019.

Bulls:

(T) = Thonglaw 1947–1974

(P) = Packy 1962–2017

(H) = Hugo 1960–2003

(T) = Tunga 1966–1998

ELEPHANT
Barn

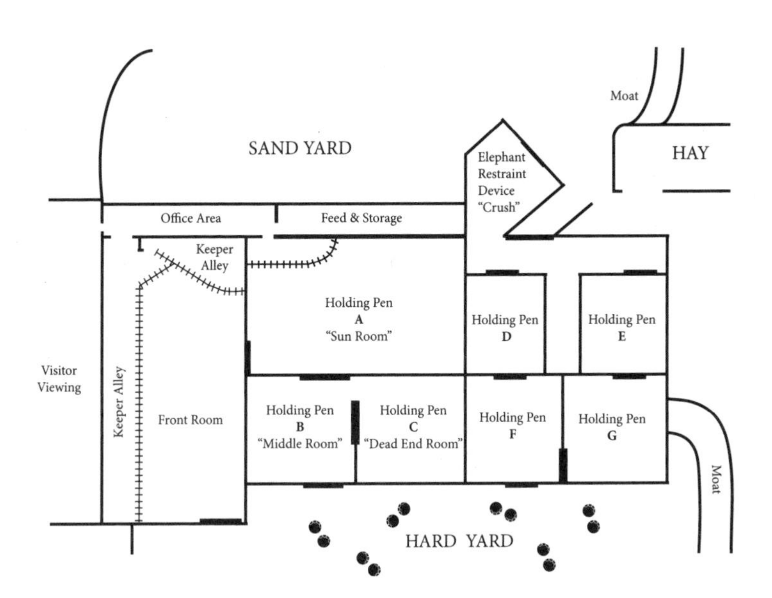

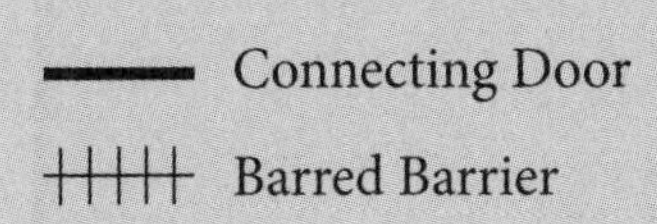

Front Room:

- Hanako's attack
- Calves are born
- Me-Tu accident
- Rosy charges Roger

Holding Pen A:

- Belle's surgery

Holding Pen B:

- Me-Tu runs Jay Haight up the wall
- Roger breaks shoulder

Holding Pen E:

- Thonglaw dies

Holding Pens F and G:

- Packy bites off Hugo's trunk

Prologue

Thursday, March 20, 1997

The Oregon Zoo began to settle down as the last of the day's visitors departed. The metal gates were swung closed and locked for the night. Staff completed their duties and prepared to head home. Nocturnal creatures stirred, ready to forage, while those who held to a daylight schedule bedded down and prepared to sleep. Animal keepers, those members of the zoo family most envied by visitors, took a final stroll through exhibits to ensure that each animal was safe and where it was meant to be, every latch secure, every door locked.

Inside the elephant barn, sixty-year-old Roger Henneous walked from room to room, boot soles almost silent against the concrete floor, his close-cut, graying beard and canny eyes shadowed by the water-stained brim of his trademark campaign hat. A veteran employee of nearly thirty years, most of them spent as senior keeper to the elephants, Roger knew every inch of the building, every sound and sigh made by the animals around which he'd built his life. The smell of timothy hay, grain, and the dense, somewhat sweet musk of the elephants washed over him, a mélange of odors so familiar as to go unnoticed. Confident his crew had handled the afternoon chores, he nevertheless made an almost unconscious mental note of each elephant as he strolled past: the bulls Packy, Rama, and Hugo, each in his individual bachelor's quarters; Pet and Hanako in their little herd; and Sung-Surin with the orphaned Rose-Tu sticking close, the teenage cow providing comfort to the two-and-a-half-year-old calf in the absence of their matriarch, Belle.

By day, the barn's cavernous space echoed with the hum of immense hydraulic doors opening and closing; the scrape of shovels and rakes clearing away manure, old bedding, and leftover bits of forage made useless with urine; the tumble of grocery produce tipped from containers; the soft thud as hay bales were tossed to waiting elephants; the sound of water spraying from hoses; and beneath it all the voices of the keepers rising and falling by turns as they talked, laughed, griped, and cajoled. And, of course, there were the elephants—squeaking, squealing, rumbling, roaring, chirping, trumpeting, and barking. Even in sleep, they broke the nighttime silence with snores and farts.

Roger stepped into the keeper alley near his office and looked into what was colloquially known as the front room, a rectangular exhibit area with bars along two walls. A temporary barricade of linked chain had been strung across its width, dividing the almost 1,400 square foot space into a smaller convalescent ward for Belle.

She stood facing the back wall, seemingly unaware of his arrival. Her left front foot was wrapped in a thick bandage secured with gray duct tape, evidence of yesterday's surgery. It was a heartfelt attempt on the part of a gigantic crew of devoted helpers to halt the advance of severe pododermatitis, known in keeper parlance as foot rot.

Social by nature, elephants prefer the company of their own kind. Belle was in isolation to protect her during this crucial recovery period. If one of the more obstreperous cows were to challenge her for the matriarchy, Belle could be severely injured. It had happened before, in 1983—Belle bested the upstart, but the elbow in her right front leg was damaged and she'd never regained full use of the joint.

To see any of his elephants withdrawn and in pain caused Roger distress, but Belle was his personal favorite, his darling. They had a special connection, a bond forged by time and shared experience. So much of a keeper's relationship with an animal was telepathic; they either clicked or they didn't. Each of the elephants in his barn had their favorite human, the one they would do anything for. Roger was that keeper for Belle.

Belle gained notoriety in 1962 when she gave birth to Packy, the first elephant born in the Western Hemisphere in forty-four years. Packy's birth heralded Portland's ascent to "Elephant Capital of the World," but Roger loved Belle for different reasons: her affectionate nature; her talent for managing other elephants; her willingness to work with keepers and veterinarians; and how hard she strove to understand these strange, two-legged creatures into whose care she'd fallen. To see her now, like this, broke his heart.

Removing a small apple from his pants pocket, he angled his slender body between the iron bars and softly spoke her name, his voice rough with smoker's gravel despite not having a cigarette in years. One ear twitched in his direction and Belle's trunk lifted, curled in an S shape to scent the air as he approached. Dwarfed by her size, he palmed the fruit into her mouth, talking in a low, soothing tone as she half-heartedly chewed. Pulp and juice dribbled from her lips to spatter the floor.

He reached behind Belle's ear and gave it a tickle. "I'll be back, darlin'," he said.

In the visitor viewing area outside Belle's enclosure sat two metal folding chairs to accommodate the keepers, and any volunteers, involved in the round-the-clock watch on their patient. Their duty was to observe Belle for signs of distress, the presence of blood or mucus in her urine or stool, and whether she ate or drank. The volunteers' intentions were good, but only Roger and his team knew the subtle clues to look for, those things they'd learned over years of working with elephants.

Roger sat in one of the chairs, balanced on the edge of his seat, elbows on knees, gnarled hands clasped. Now and then, he glanced at his wristwatch, noting the passage of time. The ball of his thumb brushed across the scratched surface of the Timex and he thought of Me-Tu—who'd once broken it—and of Rosy, and all the other elephants that had blessed his life. The barn held so many memories. On nights like this, it was easy to feel the presence of old friends, two-legged as well as four.

After an hour of quiet watching, Roger stood and left the room, returning a few minutes later lugging a large plastic garbage can. Bright green fronds of bamboo sprouted from the top. Belle raised her head at his call and slowly shuffled around to meet him. Reaching into the bin, he offered her a carrot, but she displayed no interest. He dropped it and brought up a mixed handful of hay and bamboo. This she accepted, grinding the offering between her immense molars. He dug deeper, searching for something tempting, and found a banana. She took this as well and chewed slowly, peel and all, but refused further handouts. He put the bin away and dragged out a long black hose. Placing the nozzle in Belle's mouth, he let her drink her fill. Once again, he stroked the back of her ear and promised to return. He was glad she'd eaten, but sensed her compliance had less to do with the state of her appetite than with a general desire to please him.

Roger sighed as he settled back onto his chair in the visitor area. It had been a long road to this night, but he couldn't imagine being anywhere else.

Chapter *One*

BEFORE THERE WERE ELEPHANTS

1937–1962

Three-year-old Roger Henneous slid his hand into the warm darkness beneath the last hen and drew out an egg. The chicken glared with a beady, unblinking yellow eye and jabbed at him with her beak. Snatching back his hand just in time, he carefully placed the egg in his basket along with the others, then gave the hen a retaliatory shove beneath her tail, sending her to the floor of the coop in a feathery flop. As he reached for the last two eggs, the bird struck again, pecking his legs through the denim of his overalls as she clucked a storm of birdish invective.

Roger kicked at her and swore in return, displaying verbal versatility unexpected in a child his age. He'd learned the words by following his dad around the family farm, but even at three the boy had sense enough to keep his voice low so his mom wouldn't hear. Backing away from the attack, boots flailing, Roger exited the coop and slammed the door in the chicken's face. Beneath an Iowa sky washed red with the rising

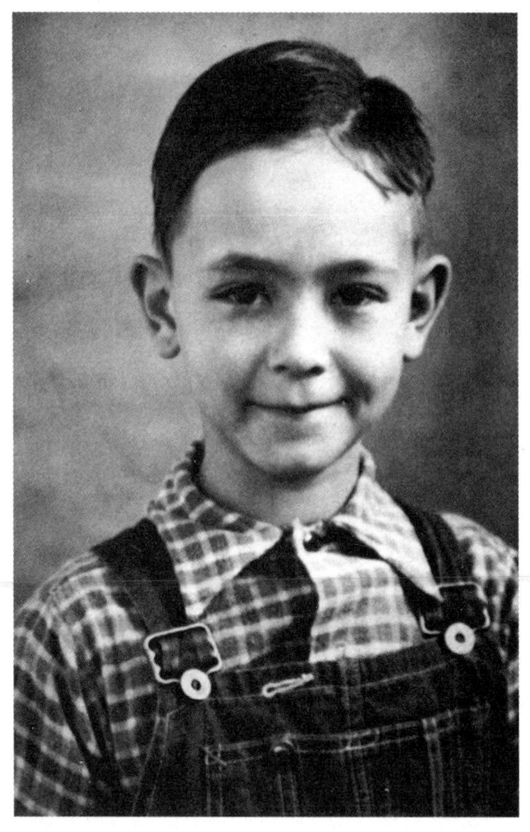

Portrait of young Roger Henneous. (Personal photo, 1945)

sun, he hurried across the farm yard toward the house where his mother waited with a hot iron skillet to begin breakfast.

As children of the Great Depression, Roger and his younger sister Virginia learned early the value of gainful employment and the obligation they owed the family to do their share. Leonard and Myra Henneous were rarely idle and expected the same of their children. Roger's introduction to chores began with gathering eggs and stacking dried corn cobs for use as fuel in the kitchen stove. By the time he entered first grade, he was in the barn before daylight helping his dad milk cows.

If the little family didn't exactly thrive, they at least endured, buoyed by the presence of nearby relatives willing to lend a hand or lighten a burden. Leonard had four brothers, two sisters having died in infancy, and Myra was one of twelve children (three of her sisters also died young). The two clans formed a rambunctious conglomerate whose shifting alliances among the women kept tempers at a constant low boil.

Small for his age and a tad scrawny, all elbows, knees, and jug-handle ears, Roger was frequently teased by his classmates and often assaulted by bullies. Barely a week went by that he didn't drag home with a black eye or a split lip, the knees of his pants torn out in the scuffle, which Myra patiently mended by lantern light after dinner. After watching his nephew wind up on the losing end of too many battles, Myra's brother George decided it was time for Roger to learn the art of self-defense. His method was simple: Figure out who the main bully was, challenge him to a fight, then get in your licks and don't back down. That way, win or lose, the other kids would know that they might be able to knock Roger down, but he wouldn't stay there for long.

Roger's true solace lay in roaming the countryside with his dog, Laddie. In the fields and woods around Blencoe, Iowa, they played cowboys and Indians, hunted for squirrels, and fished the waters of the Missouri River. On hot days, they'd pilfer a watermelon from a neighbor's field, carry the armload of dense fruit to a hideaway, and gorge on the sweet, succulent, sun-warmed red flesh. Afterward, a leisurely swim washed off any evidence.

When the United States entered World War II, every man in town beat a hasty path to the recruitment office. As "head of household," Leonard was offered a military exemption, but he declined, feeling that it was cowardly to accept an easy out when so many of the men he knew would soon be risking their lives. Roger's mother never entirely forgave this altruism, but

Leonard preferred to live with her displeasure than deny his duty. Before leaving home, he made seven-year-old Roger man of the house and told him to take care of Myra and Virginia.

Roger possessed a child's rudimentary idea of war, his knowledge based mostly on snippets of overheard adult conversations and radio broadcasts. His daydreams were full of images of his father heroically leaping out of trenches and charging into battle, or rolling through a city in a tank, or patrolling the sky as a fighter pilot. It never occurred to him that Leonard could be wounded or killed. Roger's father was his hero. As such, he was indestructible.

Despite the long odds, every man in Roger's family that enlisted came home, but they returned forever changed by the horror of the battlefield. Funny and affectionate Uncle George suffered from night terrors and took to hiding in a closet whenever an electrical storm swept through. Leonard's usual quiet reserve developed into a deep internal silence, a well of dark depression from which unexpected fits of withering criticism or uncontrolled anger flashed at the least provocation.

What was then called combat stress reaction, battle fatigue, or shell shock is now better known as post-traumatic stress disorder. In those days, the condition carried a mark of shame, as if those who suffered from it were weak, cowardly, and lacked moral fiber. No treatment existed other than for those affected and their families to live with it as best they could.

Roger didn't understand why his dad changed, but he quickly learned to deflect the brunt of Leonard's outbursts from Myra and Virginia onto his own narrow shoulders, and it wasn't long before the ridicule and rage undermined his confidence and self-esteem. Difficulties in school only made things worse. Never a strong reader, he floundered and to his shame was held back to repeat a grade. If not for the intervention of his mother's sister Dorothy, he might have eventually

Young Roger with his favorite playmate, Laddie. (Personal photo, 1947)

dropped out of school. Sensing that her nephew would enjoy something more exciting than a textbook, she gave him a copy of *Dave Dawson at Dunkirk* by R. Sidney Bowen. Roger struggled through it page by laborious page, lips moving silently as he sounded out each word. In the end, he enjoyed the book so much, Dorothy gave him another, not bothering to reveal there were fifteen titles in the series. By the time he'd reached the end, reading was his favorite pastime.

With Leonard home, the family turned to sharecropping and raising animals for their meat, eggs, and milk. In winter, when the fields lay fallow, Leonard supplemented their income by laboring in the fetid meat-packing plants of Sioux City. Before and after school, Roger helped his father work the farm and care for the livestock, while Virginia assisted their mother with household tasks.

In 1947, Myra bore a son, Donald Henneous, who quickly became ten-year-old Roger's biggest fan. As soon as he could toddle, Donald followed his big brother everywhere, keeping up an endless barrage of commentary and questions that amused and annoyed Roger by turns.

In February 1950, the leather plunger on the farm's well pump cracked. The house and barn had no indoor plumbing, so water was drawn one heavy bucketful at a time from a manual pump attached to a well pipe located in a five-foot-deep by four-foot-square pit packed with straw to keep the works from freezing. Until the pump was repaired, everyone on the farm, including the animals, would go thirsty.

Leonard and thirteen-year-old Roger suited up in layers of wool and flannel and headed outdoors, determined to have the work done before nightfall. They successfully separated the pump from the well coupling and installed the new plunger, but when they tried to screw the pump's four-foot pipe back onto the coupling, it cross-threaded every time.

The temperature hovered in the single digits. Frigid air turned their breath to steam and gnawed the naked skin around their eyes, exposed above the scarf line. Down in the pit, Roger could barely feel his hands and feet. Leonard was equally miserable standing on the rim to guide the pump from above, but there was no going indoors until the job was finished, and evening chores were waiting.

Tired and frustrated, Roger flexed his stiff fingers inside their gloves and lined up the two pieces of metal for the umpteenth time. Once again, the threads skewed. Leonard's foggy profanity sliced the air. What the hell was wrong with Roger that he couldn't make the simple connection? Why didn't he pay attention and do it right? He was too slow, too weak, too this, too that.

Roger threw down his tools and clambered up out of the hole, squaring off against his dad for the first time in his life. "You're so goddamn smart! Why don't you go down there and let me guide it instead? Maybe then we'll do better."

Outraged, Leonard grabbed him by the coat and hurled him into the pit. The upright well pipe whistled past Roger's face as he fell. He landed hard, breathing heavily, stunned less by the fall and Leonard's rage than by how close he'd come to being impaled. After a moment, he looked up at the silent figure of his dad standing above, silhouetted against a sunset sky.

"Let's try it again," he said, and climbed to his feet.

At last they succeeded and closed up the well. Roger put away his tools and immediately walked to the barn to commence milking. Seated on a stool with his bare hands full of warm teat and his forehead pressed against a bovine-scented flank, he briefly directed a stream of fresh creamy milk into a barn cat's waiting mouth.

From somewhere nearby, Leonard cleared his throat. "I'm sorry," he said. "Jesus Christ, I could have killed you. I swear to God I'll never do anything like that again."

Roger's hands paused a fraction, then resumed their work. Glancing beneath the cow's belly, he spied his dad's work boots where Leonard sat milking at the next stanchion. "You didn't mean it. I know that."

"Your mother won't see it that way."

"Does Mom have to know?"

Roger's willingness to forgive the incident and not divulge it to Myra made a deep impression on Leonard. Things between them eased somewhat after that, but Leonard's uncontrollable episodes continued to exact an emotional toll on the family. Roger grew adept at keeping his feelings submerged, but grieved for the one thing he wanted most and could never seem to achieve: his father's pride.

That spring, Roger fell for a new girl in town, a two-year-old bay mare of mixed ancestry named Lady. The young neighbor who'd received the green-broke, headstrong horse as a gift had no interest in riding and was perfectly happy to sell her for twelve dollars. A huge fan of the Old West, Roger rode Lady everywhere and soon had her trained to herd cattle.

During a particularly dry summer, the family owned more cattle than they had pasture, so Roger put them out to graze beside a roadside ditch while he watched from horseback. Along the fence line, there was an area of washed-out ground where he could ride Lady beneath the bottom wire if he bent low against her neck. One day, as they passed through and she lunged up the opposite slope, the saddle horn caught beneath his sternum and instantly dislocated six ribs.

The pain was indescribable, white-hot and sharp. Each step Lady took felt like knives driven into his chest. Unable to ride and barely able to breathe, Roger dismounted. He tied his wrists together with the reins and looped them over the saddle horn, which lifted his ribcage and eased the agony somewhat. Slowly, he began to walk. Lady matched his steps and never faltered, spooked, or stopped to graze. They reached home and Myra came running. Roger refused to go see the doctor until the mare was settled in her stall. It was weeks before he could ride again, but he never missed a day visiting Lady to thank her for helping him.

His affinity for animals didn't go unnoticed in the family and there was much speculation about whether he would attend veterinary school. Roger yearned to go, but a lack of confidence kept him from pursuing college at all. Roger always assumed he would make farming his life, but after almost two decades of watching his parents struggle against the vagaries of weather and market rate, he preferred something with fewer gambles that would guarantee a living wage.

Leonard and Myra had reached the same conclusion. In June 1956, they announced plans to relocate to Portland, Oregon, following a trail toward greener pastures laid down by Aunt Dorothy and her family. Ten-year-old Donald would go with them, but Virginia planned to remain behind, living with friends until she graduated high school the following year.

With his family about to be scattered, it took no arm-twisting to convince Roger to accompany high school buddy Bernie Carroll on a trip to California to seek their fame and fortune. In January 1957, carrying two suitcases and their entire life savings—about $400 apiece, nearly $3,600 by today's standards—they set out in Roger's Chevy Coupe, thrilled at the romance of the open road and the adventure ahead.

Expenses were small. Gasoline cost just twenty-three cents a gallon, and thirty-seven cents purchased a loaf of bread and a jar of grape jelly. When Roger and Bernie wanted to splurge, three dollars bought two chicken dinners and a six-pack of beer. They washed in gas station restrooms, slept parked by the side of the road, and opted for a motel room only when the need for a shower or a real mattress became desperate.

In Arizona, a sign advertising the North Rim of the Grand Canyon drew them off the highway on a long detour. They arrived to find the canyon fogged in, the magnificent view obliterated by a swirling wall of white mist. Deeply disappointed, the two friends solemnly performed a commemorative urination into the abyss and continued on their way.

California was kind to Bernie, who had family living there, but proved to be no golden land for Roger. After two

Twenty-one-year-old Roger in the port of Argentia, Newfoundland. (Personal photo, 1958)

unsuccessful weeks of searching for work, he decided to head for Portland, hopeful the city would be as good to him as it had been to his family. Each morning, he left his parents' home and trod the streets in shoes damp from the day before. Every evening, he returned with nothing to show for his efforts. He was giving serious consideration to farming again, no matter how dismal the pay, when his cousin Ralph called from Iowa. He was in the same predicament as Roger, eager to work but unable to find employment, and had just discovered the Coast Guard offered training, adventure, and best of all, a paycheck. Within the week, the cousins were reunited and had taken their oath of enlistment. A few days later, they arrived at boot camp in Cape May, New Jersey, the farthest east either of them had ever traveled.

The first night in barracks proved to be a mind-altering experience for the rustic, somewhat dewy-eyed pair. Crammed together with street-wise, smart-mouthed, half-grown boys from metropolitan areas like New York City, Boston, and Philadelphia (some of whom accepted enlistment over incarceration), they listened in wide-eyed wonder to lurid tales of family fistfights, hard liquor, and fast women. Taunted for his Midwestern accent and naive, squeaky-clean demeanor, Roger countered by giving as good as he got. His gumption and quick wit made him an instant favorite despite his willingness to follow orders and his ability to roust out of bed bright-eyed at the first notes of "Reveille."

Eight weeks later, Roger departed for Mechanics School in Groton, Connecticut. In September 1957, the newly graduated twenty-year-old diesel mechanic left New England for his first ship, the USCGC *Chincoteague*, a converted 310-foot former Navy aircraft fueling vessel homebased in Norfolk, Virginia and headed to Ocean Station Bravo in the North Atlantic.

The Iowa landlubber took to seafaring life surprisingly well. He enjoyed caring for the ship's engines and recording pressure and temperature readings in the shaft alley, often while balanced on the bulkhead instead of the deck when the ship rolled in heavy seas.

In 1959, he transferred to USCGC *Balsam*, a 180-foot buoy-tender out of Ketchikan, Alaska. To Roger's great delight, he discovered a dog living on board. No one knew the origins of Turnbuckle the beagle, but he was treated as a member of the crew. Wearing a collar printed with his serial number and home station, "Turnie" kept the men company while on duty and even joined them for shore leave, where he enthusiastically courted the local female dogs. When his late-night trysts kept him from catching the ship leaving pier, he'd stroll over to Base Affairs Quarters and bunk down with some kind soul until the Balsam returned. The dog's presence was a slice of heaven for Roger, and he never passed up an opportunity to give Turnie a pat or a treat.

Roger aboard the USCGC *Balsam*, 1959. (Personal photo, 1959)

Roger spent his final six months of duty at the Coast Guard air station in Port Angeles, Washington, elbow-deep in engines at the Boats and Vehicles Shop. He registered his Honorable Discharge from the United States Armed Services with the State of Oregon on February 27, 1961, settled in at his parents' home in Portland, and immediately set out to look for work.

His hunt for employment continued as the winds of March became April's gentle rain, transforming into May flowers. Day after day, as he searched without success, a safety net lurked at the back of his mind: a hefty bonus of six month's pay if he re-enlisted in the Coast Guard within ninety days of separation. Roger had no burning desire to make the Coast Guard a career, but if he had to re-enlist, he intended to take full advantage of the offer. As June roses began to bud and bloom, delighting the senses with their vibrant colors and heady aroma, Roger resigned himself to the inevitable and packed his sea bag.

Fortune unexpectedly smiled on day eighty-eight. The job at Freightliner Truck Company wasn't his first choice, but it was a beginning and that was all he'd ever wanted. What he did with it was up to him. Gleeful, he donated his uniforms to Goodwill and drew a deep, cleansing breath. For the first time in a long while, Roger Henneous felt like a free man.

Chapter *Two*

PACKY

May 1962

Waiting in line with his parents to pay admission to the Oregon Zoo, Roger gazed around with bewildered amusement at the jostling, chattering throng. He'd never expected to encounter a crowd this size, though the line of cars backed-up for miles, creeping its way toward the exit for the zoo, should have provided a clue. The parking lot had been no better. Traffic jammed the roadways, horns blaring as drivers glared through their windshields at one another, vying for the next empty space. Wiser, more practical folks parked at the far end of the lot and walked to the entrance. Others risked a fine by leaving their vehicles on the grassy verge. One bus after another pulled up to the curb and discharged passengers who immediately got in line to buy a ticket. It seemed like half the civilized world was landing here—and all because of a single little elephant named Packy.

They stepped through the gate and were immediately swept up by a rushing tide of visitors headed toward the far end of the zoo. In the distance, Roger spied a blue flag announcing

"It's a boy!" fluttering from a roof peak. The crowd saw it too and surged forward.

***

Pharmacist Richard B. Knight likely had no inkling his actions would make him the father of the Oregon Zoo. He was just looking to maintain domestic bliss.

The Englishman settled in Portland in 1882 and opened a store at 72 Morrison Street, not far from the shipping docks along the Willamette River. Lauded for his generosity (he often filled prescriptions for families that could not afford to pay), he earned a reputation as someone willing to purchase the parakeets, monkeys, and other small exotic fauna brought home by sailors from various ports of call. Although Knight reportedly had no particular fondness for animals, he collected the creatures into a menagerie and housed it at the rear of his pharmacy.

Mrs. Knight apparently had no more affection for animals than did her husband, but she tolerated their presence and the expense of keeping them until 1888, when Knight paid $125 (about $3,312 today) for a male brown bear and a pregnant female grizzly he kept staked in a vacant lot next door. Justifiably concerned the bears' proximity put their children at risk, Mrs. Knight insisted he get rid of them.

In June of that year, Knight petitioned then Portland Mayor Van B. DeLashmutt and the City Council to lodge the bears at the city park. He received a counter-proposal: two old circus cages to house the bears, and a plot of land in City Park (now Washington Park) on which to display them, with the understanding that he and his family remained responsible for the animals' welfare. Not long after, Knight made the city a second offer: they could have the grizzly and her cub for free (no mention is made of the brown bear). The mayor and council accepted, and the Oregon Zoo was born.

Park groundskeeper Charles Myers, a gardener and florist from Germany, became the bears' caretaker and created what was, for its time, an innovative exhibit: a bar-less natural area with a sunken grotto. By 1906, the zoo had expanded to more than 300 species, including a seal, deer, alligators, kangaroos, a polar bear, and a lion. The number of visitors increased, but overall the new century proved unkind to the zoo. City funding declined, impacting animal care and building maintenance. A 1909 newspaper article decrying all zoos as inhumane so incensed readers, they wrote to the city demanding its closure. Despite these difficulties and others, including an escaped lion, infestations of rats and earwigs, and utter lack of interest on the part of park commissioners, the zoo remained a popular destination for citizens and visitors.

After World War I and the subsequent economic downturn, the zoo was moved to a remote part of Washington Park (now the site of the Japanese Gardens) where it languished for twenty-five years. At the commencement of World War II, funding was cut even further. When the war ended, the necessity to create jobs prompted renewed interest. But as multiple bond measures failed over a three year period, it seemed the dream to revitalize the zoo was doomed.

Then something miraculous occurred. In 1954, Rosy came to town.

The four-year-old female Asian elephant, given to Portlanders Catharine and Austin Flegel in gratitude for Mr. Flegel's role as Chief of the U.S. Technical and Economic Mission to Thailand, may seem an unconventional present, but elephants have been given in tribute, most often between rulers, since Harun ar-Rashid made a gift of the Asian elephant Abul-Abbas to Charlemagne in 802. There have been several such grand gestures throughout history. In 1861, President Abraham Lincoln received an offer of several mating pairs of elephants from the King of Siam, but respectfully declined the gift.

The Flegels christened their elephant "Rosy," in honor of Portland's fame as "The City of Roses," and donated her to the zoo. She traveled by train from Chiang Mai to Bangkok, where she boarded the SS *Washington* bound for the United States, transferring to the SS *Montana* for her journey to Oregon. On September 14, 1953, Mayor Fred Peterson welcomed Rosy to her new home before an immense crowd of cheering well-wishers. A children's book, *ROSIE: the Elephant at the Portland Zoo*, was published in her honor, and her social calendar filled with events: parades, store openings, even an opportunity to throw out the first ball at a Portland Beavers baseball game. She was gone so often from her enclosure at the zoo, visitors complained.

In the wild, elephants live in large matrilineal herds composed of grandmothers, mothers, aunts, sisters, and calves of either sex. They should never be housed in isolation, but in the early 1950s few people understood that, or much of anything else about the species. Rosy must have been very lonely

Rosy is ceremoniously welcomed to Portland, Oregon, 1953. (City of Portland (OR) Archives, A2005-005.1330.12)

living on her own in the zoo's dilapidated camel barn. She had plenty of human visitors, but none of her own kind, and no dedicated keeper until Alvin Tucker arrived in 1954. Originally from Kansas, he'd traveled west to work in the Portland shipyards during World War II. When peace made him superfluous, Tucker found employment with the City Park Bureau and was assigned the task of looking after Rosy.

Quiet and thoughtful, watchful, empathic and perceptive, Tucker took his job seriously. Rather than force his will on the young elephant, he allowed Rosy to tell him what she needed. He learned to read her body posture, the position of her ears, and her vocalizations. He contacted circuses and other zoos to learn how they handled their elephants, then sifted through the information, discarding what he didn't agree with, and putting to use what worked for Rosy. As the bonds of trust and affection strengthened between them, she responded to his presence, greeting his arrival each day with squeaks and squeals of delight. Meanwhile, her popularity grew, and voters finally approved funding for a new zoo. Construction began in 1955.

The zoo received its second elephant the following year, a gift to the children of Portland from the King of Thailand. Originally named "Ike," in honor of President Dwight D. Eisenhower, and later re-christened "Buddy," the four-year-old male received an enthusiastic welcome from Rosy and Tucker. Four months later, a Portland engineer named Orville Hosmer donated a third elephant, a year-old female he'd received as a gift for his help rebuilding a Vietnamese village destroyed by flood.

Tuy Hoa (pronounced *Tee-Wah*, meaning "peace") was terrified, disoriented, and much too young to be separated from her mother. Tucker had no idea how to console the distraught infant, but Rosy stepped into the role of surrogate mother as though born to it, a hint that she'd likely received some training from her own mother before being captured and sold. She wrapped her trunk around the trembling calf, drew Tuy Hoa close to her side, and crooned. From that moment on, the two were inseparable.

Within months of Tuy Hoa's arrival, it became obvious something was seriously wrong with Buddy. The little bull's legs bowed so badly, he had increased trouble standing and moving around. Examinations by two different veterinarians brought slightly different diagnoses—one thought deformity in the hip joint, the other pelvic malformation—but the prognosis was the same: the condition was progressive with no hope of reversal or permanent relief.

Buddy was euthanized in March 1957. Tucker mourned alongside Rosy and Tuy Hoa, and wondered what he might have done differently to help the bull. A present-day veterinarian has suggested that if Buddy was hand-raised—as seems likely since his donation paperwork describes him as "well trained"—he may not have received proper nutrition during his first couple of years. Metabolic Bone Disease, also known as glass-bone disease, is another possibility, but cannot be diagnosed without radiographs or blood work, neither of which exist.

Rosy and Tuy Hoa moved to the new elephant barn in 1959. In the autumn of 1961, they were joined by Belle, Thonglaw (pronounced *Tung-Law*), and Pet, three Asian elephants belonging to animal trainer and importer Morgan Berry. A classical musician, Berry became interested in exotic animals while playing in a band on cruise ships. He and his elephants performed at Seattle's Woodland Park Zoo during the warmer months of the year, and now Berry had arranged a deal with his friend, Oregon Zoo Director Jack Marks, to put them on exhibit in Portland during the winter to allow him to travel overseas and purchase more animals.

Overnight, Tucker's duties increased exponentially, as did the personalities he dealt with. Rosy, Buddy, and Tuy Hoa taught him no two elephants were alike. With the arrival of Berry's herd, his education ramped up significantly.

Thonglaw, a majestic and powerful fourteen-year-old with impressive tusks, was declared out-of-bounds, controllable by no one save Berry and his son Kenny. Nine-year-old Belle had been hand-raised from infancy by Berry's family. Entirely acclimated to humans of all shapes and sizes, she possessed an innate ability to control her own kind. She and twelve-year-old Rosy learned to respect each other as one matriarch to another, but never became close, and both held definite opinions on how things ought to run in the barn. Big-headed, four-year-old Pet was endearingly pigeon-toed and utterly devoted to Belle who, like Rosy with Tuy Hoa, acted as her surrogate mother.

What no one knew at the time, except for the elephants, was that Belle was pregnant. Berry suspected, but it wasn't until the winter of 1961–62 that a fetal EKG confirmed the presence of a calf. No one knew the length of an elephant pregnancy, and there were several false starts over the ensuing months when Belle's nervous handlers (Berry, Tucker, Marks, veterinarian Matt Maberry, and others) thought she was going into labor. Marks' anticipation of the event so stressed him that when Belle finally delivered on Saturday, April 14, 1962, he promptly passed out. In the furor of labor and delivery, no one thought to alert Al Tucker at home and the senior keeper missed the entire event. Belle celebrated the arrival of her son by consuming a bouquet of flowers, and Papa Thonglaw ate a cigar.

Newspapers headlined the birth and songs were written in celebration. Over 3,000 people entered radio station KPOJ's contest to name the calf, and school teacher Wayne W. French triumphed over such offerings as Belle-Boy, Ding-Dong, Porty, and Nogero (Oregon spelled backward). LIFE Magazine

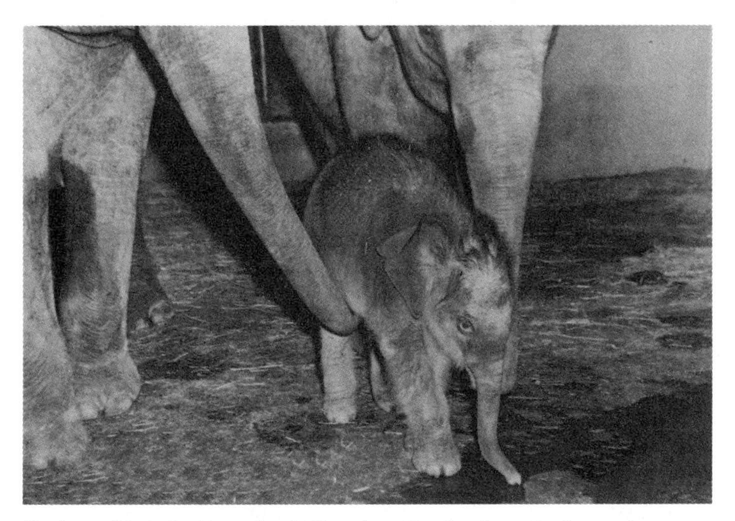

Newborn, "Packy," with mother Belle and another female on April 14, 1962. (Oregon Historical Society, 1962)

devoted an unprecedented eleven pages to the event. Author Aldous Huxley and his brother, eminent biologist Sir Julian Huxley, were among the celebrities to welcome the newcomer.

Overnight, Packy changed the face of Portland forever.

***

The pedestrian momentum carried Roger and his parents all the way to the pink-painted elephant barn, where they washed up at the end of a long, switch-backed queue leading to an open doorway. Within moments of their arrival, hundreds of people fell into line behind them.

As they inched ahead, shifting from one tired foot to the other, chatting with their neighbors to pass the time, Roger learned Belle's closely monitored pregnancy had lasted an eye-popping twenty-one months. Now a month old, the calf had gained almost five inches in height and nearly doubled his birth weight of 225 pounds.

After an hour, Roger and his parents reached the barn's threshold. Cards and gifts, including pacifiers and crocheted baby blankets, filled the space between the exhibit rail and the glass, along with hundreds of congratulatory bouquets of flowers, some wilting for lack of water.

Perched on his toes, Roger caught his first surprising glimpse of the infant and was instantly smitten. With his baggy skin and wooly reddish hair, Packy looked more like a miniature mammoth than an elephant. He toddled around Belle's legs as aunts Rosy, Tuy Hoa, and Pet looked on with obvious delight and reached to touch him with their trunks each time he passed. Every so often, he squealed like a leaky balloon and lifted his mouth to suckle briefly from the incredibly human-looking mammaries located behind Belle's front legs.

His eight-inch proboscis wriggled and twitched like a separate creature, a confusing bit of anatomy it would take the calf months to learn to control. Fascinated by the movement of Belle's trunk, Roger followed its graceful length up to her face and met her eyes through the glass.

He'd worked with enough animals to know how bright some can be, but never had he confronted such a powerful sense of intelligence. There was more than hard-wired instinct behind these eyes. Here was a brain akin to his own, but at the same time wonderfully alien. Suddenly, all he wanted was to understand it.

"Move along, please," a uniformed zoo employee called out. "We've got a long line today, folks, and want to make sure everyone gets their chance."

Roger reluctantly followed his parents toward the exit. Just before he stepped through into daylight, he craned his neck back for a final look. As if waiting for it, Packy waved goodbye with his trunk.

Chapter *Three*

THE ZOO BECKONS

1967–1968

Moving his arm in a controlled arc, Roger plied a layer of bright color from the paint sprayer across the chassis of the semi. The factory was in motion all around him, a clangorous barrage of machinery and roaring ventilation fans that assaulted his ears despite wearing hearing protection. A fine mist of airborne lacquer stippled his coveralls, gloves, boots, and any exposed skin. The particles worked their way inside his goggles and clung to his eyelashes, which he carefully cleaned with paint thinner at the end of each day. The soupy mix of chemicals coated the back of his throat, and he could taste them whenever he swallowed. Coupled with the stink of fumes, the sensation left him feeling constantly nauseated.

His wife, RoseMerrie, had done what she could to help by sewing tight-fitting hoods with wide collars that he wore tucked inside his coveralls. By sealing the wrist and ankle cuffs with duct tape and applying a layer of grease to his face, he kept the paint at bay as much as possible, but the job at Freightliner remained a daily dose of hell.

All he'd ever wanted was to work outdoors with animals, yet here he was, cooped up away from light and air, hardly seeing a living creature apart from the gaunt, vermin-infested feral cats that haunted the property. The job at Freightliner had come along when he most needed it, but six years later he felt like a prisoner.

He worked through lunch as usual with his sandwich in one hand, the sprayer in the other, and only took a break to step outside for a cigarette. Hot summer sun beat against his bare head as he smoked and stared into the middle distance, seeing not the factory lot spread out around him, but RoseMerrie's earnest expression the other night. Having grown tired of listening to him complain about his work, she'd finally issued an ultimatum: "Find a new job or find a new wife."

Two and a half years of marriage had taught him that she meant it. Unwilling to lose the woman he adored for a job he hated, Roger knew he had to find something different. A single man could just walk away and trust in fate to offer up another opportunity, but he had a wife and six-month-old daughter to support. The notion of leaving a well-paying senior position just because he didn't like the work rankled his staunch Midwestern work ethic. Employment was meant to pay the bills and provide security into the future, not serve as entertainment. He couldn't just walk away, but there were days when he felt like he'd die if he didn't.

Roger had been two years out of the Coast Guard and living with his parents in exchange for sweat-equity when his sister set him up on a blind date with twenty-four-year-old divorcée RoseMerrie Peterson. Their attraction was mutual and instantaneous, but Roger faltered in pursuing her, unable to fathom what this pretty blonde lady could possibly see in him. It wasn't until an old Coast Guard buddy wisely told him that though he might not always be happy with her, but he'd never be happy without her, that Roger worked up the courage to propose. They wed on New Year's Eve 1964, just three months after their first date, and baby Michelle arrived two years later.

Having spent a large portion of her girlhood helping to raise her younger siblings, RoseMerrie fell into motherhood without a ripple. For Roger, life with a new baby carried a steeper learning curve. He was a big brother twice over, but he'd been a farmhand and child-rearing duties were left to his mother. Now, as he watched his wife deftly juggle diapers, feedings, naps, and everything else required to properly care for an infant, he thought about his hardworking parents and his respect for them grew.

Their duty and sacrifice was much on his mind the following morning as he paged through the newspaper classifieds before heading to work. A bold-faced line of print halfway down the page caught his eye. He read it with a sense of astonishment. The Oregon Zoo was hiring a keeper. "Now there's a job I'd like," he murmured.

RoseMerrie lifted Michelle to her shoulder and patted the baby's back, eliciting a burp. "What is it?"

"Zoo keeper." He offered her a lopsided smile, ready to laugh it off.

"You should apply."

Roger snorted. "I don't have any experience with exotic animals."

"So what? You've worked with plenty of others. How different can it be?"

"Lots," he said, though he'd already begun to wonder just how much. "They're looking for people with an education, not someone like me."

"You won't know unless you try," RoseMerrie said.

Unwilling to engage in a debate he knew would escalate into an argument, Roger pushed the paper aside like a man refusing temptation and glanced at the clock. Gathering his lunchbox and thermos, he kissed his family goodbye.

RoseMerrie waited until she heard his truck drive off before she opened the newspaper to read the ad. If working at the zoo would make her husband happy and stop his constant grousing, she intended to push for it as hard as she could. Without a single twinge of guilt, she picked up the telephone and dialed.

Two days later, Roger arrived home after work to find an envelope beside his plate. He glanced at the return address—Civil Service Board—and understood at once what RoseMerrie had done. His first impulse was to pitch the application into the trash. Instead, he paused to consider. If by some wild chance the zoo did hire him, it would be a dream job so far removed from Freightliner, he couldn't imagine ever wanting to leave.

Bolstered by his wife's faith in him, and with nothing to lose, Roger completed the application and sent it off in the mail the next day. When several weeks passed without a response, he was disappointed but not surprised. He'd told RoseMerrie the zoo wouldn't be interested in him. He'd taken a chance and it hadn't panned out. Now it was time to put away the childish notion of a dream job and either resign himself to a life painting trucks or find something different.

Then, in early July 1967, Roger received a call asking if he was interested in testing for the position of animal keeper. It had taken some hapless government employee two months to wade through an overwhelming 400 applications and winnow them down to 230 suitable candidates. Had Roger realized the odds against his getting the job, he'd never have agreed, but by the time he arrived at the site and saw the crowd waiting, it was too late to back out.

He breezed through the civil service exam: one hundred questions concerning basic problem solving, simple math, and vocabulary, plus an additional sixty related specifically to animal care. On July 13, he received notice that he'd passed. As one of sixty applicants called back to participate in a physical demonstration test, he was required to lift and carry a one-hundred-pound sand bag and haul four fifty-pound gunny sacks from one place to another in a memorized routine against a time clock. In the prime of his life and strong from lugging five-gallon paint buckets, Roger again made the cut. Each of the twenty remaining candidates was granted a personal interview on July 28.

During the interview, Roger learned there was only one animal keeper position available. He left the meeting expecting to be rejected, but when it happened, he felt crushed with disappointment. He later learned his placement score—he came fifth in the list of ten finalists—had less to do with performance shortcomings than with a lack of military preference points. Disabled veterans or those who'd served on active duty during specific time periods or certain military campaigns received special consideration in hiring for federal jobs. His time in the Coast Guard worked in his favor, but four of the applicants had a greater number of accrued points.

Roger swallowed his discouragement and moved on. Months passed. He applied for a couple of different jobs around town, but didn't get those either. The family celebrated Michelle's first birthday, followed by Christmas, then Roger and RoseMerrie's third anniversary. Soon, it was spring again.

They were standing in the driveway studying Roger's truck, wondering how they were going to fit in everything they needed for vacation, when the postman came up the sidewalk and handed RoseMerrie a stack of mail. Shifting Michelle from one hip to the other, she flipped through the envelopes, checking for bills. "Roger."

Caught by the sharp tone in her voice, he turned. Wordlessly, she handed him a cheap white postcard bearing the familiar return address for the Civil Service Board. Roger flipped it over to read the back. The city was looking to hire another animal keeper and his name had drifted to the top of the list. If interested, he should check "yes" in the box below and return the card as soon as possible.

Roger hesitated, uncertain whether he could stand the disappointment if he was turned down a second time. In the end, and with RoseMerrie's urging, he checked the affirmative and added a caveat at the bottom stating that he was interested in the job only if they could offer permanent full-time work because he had a family to support. He slipped the postcard into the mailbox, squeezed half their worldly possessions into the bed of the truck, and the three of them headed north to visit RoseMerrie's grandmother in Canada.

Ten days later, their headlights swept the driveway as they pulled in after midnight. RoseMerrie sighed with relief and Roger uttered a soft, emphatic profanity by way of agreement. Michelle shrieked from her car seat, as she'd been doing for hours, her gums sore with the arrival of her first molars.

It was almost dawn by the time RoseMerrie finally soothed the baby to sleep and Roger had carried everything from the truck up to the second-floor apartment. Flat-lined with exhaustion, they shed their clothing onto the floor and crawled into bed. As sleep claimed him, Roger enjoyed a swift moment of gratitude that today was Sunday and he didn't have to work.

Three hours later, their sleep was shattered by the sound of the telephone. Narrowly quelling the desire to rip the damn thing from the wall and fling it through a window, Roger vaulted out of bed and snatched up the receiver before it could ring a second time and wake Michelle. "Henneous residence," he growled in a voice rusty with sleep.

"Good morning!" said a cheery voice on the other end. "This is Bill Scott, foreman at the zoo."

Roger's irritation evaporated. Buck naked in the middle of the bedroom, he stared at the receiver in his hand.

"Hello?" Scott said into the silence. "You still there?"

Roger cleared the gravel from his throat. "Yes."

"I know it's Sunday," said Scott, "but how soon could you come up for an interview?"

Roger's mouth went dry. "Would an hour be too long?"

Scott chuckled. "Make it easy on yourself. Let's say noon."

"Noon," Roger repeated. "I'll be there. Thank you." He hung up the phone and turned to RoseMerrie, awake and watching him. "That was the zoo," he said and sprinted for the shower, all thoughts of sleep abandoned.

Somehow, Roger made it across town without being pulled over for speeding. Scott greeted him with a smile and a handshake. When they were seated, the foreman inquired about Roger's animal experience.

"I don't have five minutes of zoological training, but I can learn," Roger said. "I've got nearly twenty years of farm experience with livestock, and near as I can tell all animals need pretty much the same things: clean water, good food, adequate shelter, and protection from people who might do them harm."

Scott explained that a lot of young people fresh out of college came to the zoo looking for work. Most seemed to think being a keeper meant Monday through Friday, forty hours a week, with holidays off. How did Roger feel about that?

He snorted. "If I learned anything during my years as a farm boy, it's that caring for livestock is a seven-days-a-week, twenty-four-hours-a-day, fifty-two-weeks-a-year proposition. Animals don't know that it's Christmas or Thanksgiving or your birthday or whatever and wouldn't give a damn if they did. They're standing in their own crap, they're hungry, they need a drink, and some need medical attention. If you're worth half a shit, you'll do those things. If you're not prepared to, then you need to get a desk job shuffling papers."

A faint smile touched Scott's mouth. "Sometimes the work requires odd hours. That going to cause problems for you at home?"

"I'm already doing shift work, so my wife's used to that."

"If you get the job, how soon can you start?"

Roger smiled. "I'd have liked to start the minute I walked through the gate, but I've got seniority with my present employer and owe them two week's notice."

He later learned that his commitment to not just walk away from his current job was one of the high-water marks of the interview as far as Scott was concerned. In the past, too many keepers had grown disenchanted with the labor involved and had quit without warning, abandoning their animals and doubling the workload for the staff that remained.

The foreman scribbled another note, put down his pen, and told Roger he would start on June 2 at 7:30 a.m. They shook hands and Scott offered his congratulations. Roger was halfway across the parking lot when his knees turned wobbly as the realization hit him. The next morning, he handed in his two weeks' notice at Freightliner.

True to his nature, Roger almost immediately began second-guessing his decision to leave his job. The keeper position meant a whopping 33 percent pay cut. Some of that money might be regained over the long haul in the form of raises, and maybe bonuses, but never all of it.

RoseMerrie didn't care. They'd both lived through rough times before and could do it again. She was determined to make whatever sacrifices were necessary to give Roger this chance.

Over the next two weeks, he called home every day to ask if she was positive the job change was a good idea. Each time, he reminded her of the loss of seniority, the pay cut, and the uncertainty the future held.

Her response was always the same. "Shut up and do it!"

Chapter *Four*

LIONS, GIRAFFES, AND THE FIRST BABY ELEPHANT

1968–1969

His first day on the new job, Roger woke before dawn, too excited to sleep. The morning routine was deceptively similar, but steering his truck toward the zoo instead of Freightliner gave him a smile he couldn't erase. Even if today turned out to be the worst of his life, it was bound to be better than another minute in that factory.

Bill Scott met him at the employee entrance and they walked to the commissary, where the foreman pointed to a large chalkboard, its surface divided into squares, each one labeled with an area of the zoo: Primates, Bears, Reptiles, Hoofstock, Elephants, Maintenance, and so forth. He explained that every new keeper began as a floater, changing duty areas daily to gain exposure and experience, so the first thing he should do each morning was check the board for their assignment.

Roger's eyes roamed the blocks. With a surreal sense of wonder he found his last name written in the box labeled

"Feline." Scott explained that house belonged to senior keeper Lorin Floessler, and Roger should go find him for orientation to the area.

Walking through the zoo before the gates had opened for the day was a rare and wonderful experience for Roger. There was no crowd to jostle him, no aimless movement from one exhibit to the next, just purposeful industry. Intent on their appointed rounds, keepers and other staff hurried past, nodding distracted greetings if they happened to catch his eye. The air trembled with sound as the animals woke and began calling for breakfast.

Roger couldn't stop smiling. Less than an hour on duty and he already felt like he belonged. Seven years at Freightliner hadn't given him a shred of that.

He found forty-six-year-old Lorin Floessler busily occupied with his cats. The dark-haired, impish, and amiable senior keeper welcomed him with a warm handshake and led him on a tour of the facility.

Built in 1959 with the advent of the new zoo, the Feline House was a cheerless cement block divided into stark daytime and night cages. In the dark decades before the impact of environmental factors on the physical and mental wellbeing of animals was understood, the Oregon Zoo, like so many others, was a dreary place of concrete, iron bars, and cramped quarters; what one former employee described as "easy-to-clean handball courts." The big cats—lions, tigers, jaguars, and leopards—were the main attraction, but the smaller felines, such as cougars and bobcats, drew their share of interest as well. This early in the day, they were all up and moving, impatient for breakfast, stomachs rumbling as they paced incessantly in their confinement, luminous eyes fixed on a horizon no one else could see.

Floessler explained the daily routine: feeding and watering the animals, noting any changes in their behavior or physical condition, dispensing medication to those that needed it, cleaning cages, and "bolt and bar" checks, a form of building maintenance that involved inspecting cage hardware, making sure that all welds are intact and all door hinges clean of rust. No matter what the duty, an eye for detail was imperative and could be lifesaving. A keeper who let his attention wander for even an instant was asking for trouble. Floessler had seen it happen more than once. Someone would stray too close to the cage bars and wind up on the receiving end of a lightning-fast strike from a paw full of claws. They'd misjudge an animal's speed or lose track of it inside an exhibit, not check a door because they "knew" it was locked, or forget to switch the "tiger in/tiger out" sign.

Having circled the building, they stepped indoors, where the sharp scent of cat intensified: a pungent, nose-tingling mix of ammonia and musk. Sleek bodies glided past cage bars with liquid grace, immense paws silent against the concrete, but Roger barely noticed, so intent was he on listening to Floessler.

Without provocation or warning, an African lion weighing more than four hundred pounds sprang at the men. The welded iron mesh enclosing his area rang with the collision and Roger teleported backwards across the alleyway. On the far side of what now seemed much too narrow a space, a pair of huge golden eyes stared unblinkingly at him, framed by a dense thicket of bristling mane.

Floessler chuckled. "Good morning, Caesar." The tawny feline rubbed his cheek against the wire. The keeper obliged by giving it a scratch.

Roger gasped, heart galloping. "Shit! Is that going to happen every day?"

"Pretty much." Floessler worked his fingers along the line of the cat's upraised jaw, smiling as Caesar's eyes closed in pleasure. "This grizzly old bastard will check your reflexes whenever he gets the chance. He'll stop once he understands you're part of the team."

Roger stared at the man and beast engaged in a communion he had absolutely no desire to share. "You didn't even flinch!"

"This big bugger has jumped at me so many times I don't have any flinches left," Floessler said, never taking his eyes from the big cat. "Anyway, the leopard is worse. A lion is slow-motion compared to a leopard. And a jaguar? They're even faster." He gave Caesar a final pat and continued down the alleyway. Roger followed, sidling along the wall until they were well past the enclosure. Caesar watched his retreat with an air of regal amusement.

"Those were pretty good reflexes, by the way," Floessler remarked over his shoulder as they walked. He grinned. "But you'd still have been too slow."

Any new keeper naïve enough to entertain romantic notions about zoo work is swiftly disabused of the idea. The bulk of animal care involves tedious, dirty labor, and lots of it. Roger returned home that afternoon filthy from head to heel, but—much to RoseMerrie's delight—happier than he'd been in a long time.

He threw himself into the work, eager to learn what anyone cared to teach and loving every minute of it. Nothing dampened his enthusiasm, not even jobs like draining and scrubbing the odious duck pond or dealing with the geese, whose snapping beaks often left welts the size of half-dollars on his legs. For $2.80 an hour (the equivalent of around $20 an hour today), Roger eagerly took on any task thrown at him, determined to do well. What he loved most was the constant variety in his tasks. Every day brought something new and interesting. It was impossible to be bored.

New hires served a six-month probation after which their performance was evaluated by a supervisory panel made up of senior personnel. Those that received high marks were retained and promoted to permanent status. The supervisors looked for initiative, common sense, skill set and the willingness to increase it, temperament, adherence to rules, and the ability to follow orders. There were those among the senior men who saved the worst jobs for their days off so the newbies would get stuck doing them, but Roger didn't care. The work needed to be done. It didn't matter by whom.

That attitude impressed a lot of people. Soon, many of the area supervisors were asking for him by name. His candor and caustic wit may have gotten him into trouble from time to time, but his unflinching work ethic made up for it. No matter how difficult the task, he stuck with it until he'd wrestled it into submission.

He never expected to wrestle a giraffe.

The zoo owned two: a big male named Polka and his female companion, Dot. Polka's hooves had overgrown to the point where they'd begun to curl up along the outer edge. In the wild, travel over great distances keeps a giraffe's hooves worn down and in shape. At the zoo, where distance options were severely limited, that job fell to the keepers. Roger had learned how to trim hooves back in Iowa, tending to his family's and his neighbors' livestock. When his skill with nippers and rasp became known among his coworkers, he immediately became the guy to call when in need.

Roger regularly tended to the goats and sheep in the Children's Zoo, but as comfortable as he was with hoofstock, a giraffe seemed a bit out of his league. Still, when asked, he joined the crew headed by Lorin Floessler's older brother, Lloyd, outside the giraffe enclosure. Behind the fence, eighteen-foot-tall Polka watched them with benign regard and slowly chewed his cud as they assessed the situation. The question wasn't whether it could be done—it had to be, or Polka would eventually be crippled—but how to do it.

Their concern was well-founded. Despite their gangly appearance, long-lashed movie starlet eyes, and comically sweet and gentle expression, giraffes are dangerous. Able to achieve bursts of speed up to thirty-five miles an hour, they can easily overtake a fleeing man, and a backward kick from one of their feet can decapitate a lion.

After some thought, Roger offered a suggestion. He'd seen his dad subdue nervous horses by tying a piece of sacking across their eyes, cutting off their line of sight to whatever scared them. If they could get a bag over Polka's head, one long enough to fit down his neck so he couldn't shake it off, maybe that would keep him calm while they trimmed his feet.

Floessler decided it was worth a try. He called the zoo's fabrication shop and ordered a bag made of dense material that was opaque, but breathable. When it arrived, they rigged it onto a pulley system over the highest feed station so Polka's neck would be at full extension when they dropped it over his head. They filled the box with tempting treats and brought the giraffe indoors.

The plan worked beautifully...until it didn't. Polka was delighted to receive a bucket of goodies, but when the bag dropped flawlessly over his head, he panicked. Thrashing his neck from side to side to dislodge whatever was attacking him, he reared up on his hind legs, poised for a second like a ballerina en pointe, and flipped over backward. His head struck the concrete wall with a horrific sound and he crumpled.

The keepers stared in horror at the inert body, then ran forward, wondering if they'd killed him. Fortunately, a giraffe's skull is naturally reinforced, a necessity in battles for mates or other desired territory. Polka had simply knocked himself out. Working quickly, they injected him with a light tranquilizer. While one group kept his head elevated so he wouldn't aspirate the contents of his stomach into his lungs and cause pneumonia, Roger and a second crew went to work on Polka's hooves. Using a carpenter's saw, they removed the hoof's overgrown edge, then switched to handheld nippers for the close work so as not to cut into the soft pulp. A rasp served to round the finished edges.

When they finished, they removed the bag and gave Polka a drug to reverse the tranquilizer. Within minutes, the giraffe was awake. Rolling onto his chest, legs tucked beneath him, he pushed to his feet. A full-body shake settled the world back into place. Satisfied by how well things had gone, Roger offered him a treat on his way out of the barn.

Roger's pleasure that day was nothing compared to what he experienced each time he was assigned to the elephant barn. Located against a hillside at the far end of the zoo and partially surrounded by a dry moat, the pink concrete building contained a front exhibit room, six interior holding pens, a yard, and a large, dusty, high-ceilinged room for hay storage that doubled as an office. To Roger, stepping through the doorway felt like coming home. The smell of grain and hay, the odor of animals, and the sounds of enterprise—doors clanging, hoses splashing, and the cheerful profanity of keepers Denslow Robbins, Joe Cochran, and senior keeper Al Tucker—gave him a feeling of peace.

Shortly after Packy's birth in April 1962, owner Morgan Berry received bids to buy the newborn calf and his mother, with one dealer offering $50,000 for the pair—equivalent to nearly $419,000 today. The threat of losing Belle and Packy spawned a wild, city-wide fundraising scramble as the citizens of Portland rallied to raise enough money to keep them at the zoo. In the end, Berry agreed to sell the pair to the Oregon Zoo for only $30,000, in part because of his friendship with director Jack Marks, but mostly because he wanted the elephants nearby so he could visit whenever he chose. In a show of generosity and a desire to keep the herd intact, Berry included Pet and Thonglaw in the deal. In May of 1962, Belle, infant Packy, Thonglaw and Pet joined Rosy and Tuy Hoa as permanent residents of the Oregon Zoo.

Much to the surprise of those who considered Packy's birth to be a fluke, the herd continued to grow, becoming an astonishing success story in captive elephant breeding. Between 1962 and 1968, five calves were born. Two remained by the time Roger arrived: six-year-old Me-Tu and five-year-old Hanako. The other three had been sold. Walk in the Wild, a Spokane zoo run by the Inland Northwest Zoological Society, bought five-year-old Dino. Chicago's Brookfield Zoo purchased three-year-old Noel and later sold her to Ringling Bros., where her name was changed to Cora. The third calf, two-year-old Rajah, was purchased by Morgan Berry and his business partner, renowned animal trainer Eloise Berchtold. They renamed him Teak and trained him to perform in Berchtold's circus act.

Although the selling of calves to circuses and other zoos was business-as-usual in those days, the practice has fallen out of favor due to what has since been learned about the importance of family among elephants. The last time an elephant calf was sold by the Oregon Zoo was in 1988 when an exchange was made with Ringling Bros. for a breeding bull.

In addition to Rosy, Tuy Hoa, Belle, Packy, Thonglaw, Pet, Me-Tu, and Hanako, the zoo housed two additional cows: Effie, on breeding loan from California's Oakland Zoo, and Winkie, from the Henry Vilas Zoo in Wisconsin where, in 1966, she dragged a three-year-old girl into her enclosure and killed her.

In Tucker's opinion, poor exhibit design was more to blame for the child's death than any independently vicious act on Winkie's part. Back then, the elephant exhibit area at the Henry Vilas Zoo was surrounded by a tall fence and separated from visitors by a fourteen-foot open area bordered by an additional three-foot decorative fence. Despite these precautions, children were routinely encouraged by their parents to slip beyond the first barrier and cross the open area to feed Winkie by hand. Jack Marks accepted her as a potential breeding elephant for the Oregon Zoo only after assuring the public the facility could keep her from becoming dangerous.

The more time Roger spent in the barn—sweeping, shoveling, washing floors and windows, and helping to prepare food—the more he came to know the elephants as individuals and recognize them on sight. Rosy, built heavy and close to the ground, had a pronounced dip in the center of her forehead easy to spot when she turned in profile. Tuy Hoa was the tallest of the cows, but more lightly built, a gangly-legged flibbertigibbet whose nervous personality could transform her internal workings from solid to liquid waste in a matter of seconds. Belle had a calm and trusting manner with humans, but retained her indomitable spirit with her own kind. Roly-poly Pet didn't have a mean bone in her body and delighted in occasionally challenging Tucker's rules. Packy, now grown from a wobbly-kneed infant to a handsome six-year-old, hated any change in his routine. Unlike Thonglaw, Packy was a *mukhna*, a tuskless bull. In Asian elephants, only males grow tusks, although not always. Tushes—a shorter version of the tusk which seldom extends beyond the upper lip—appear in both males and females of the species. In African elephants, both sexes grow tusks.

Me-Tu, nicknamed "Daughter Hog" for her love of food, behaved well with keepers, but was bossy in the herd. Hanako, whose name means "flower child" in Japanese, was the only female with tushes. She'd inherited Tuy Hoa's height and was something of a scrapper, sparring with Me-Tu or ganging up with her against the others. Effie was cranky and ill-mannered,

Hanako. (Point Defiance Zoo)

but some of that may have been due to her advanced pregnancy. Despite Winkie's violent reputation, she rarely caused a fuss and kept mainly to herself, intimidated by the combined matriarchal presence of Rosy and Belle. Roger was so in awe of Thonglaw, he had no difficulty heeding Tucker's stern warning to maintain a safe distance from the bull at all times.

***

Portland's rainy season had begun, but Sunday, September 29, 1968 was dry despite a threatening cloud cover that capped the city like a hat. Morning temperatures hovered in the low 50s, chilly enough to require a jacket, but with a high predicted to be in the 70s, Roger looked forward to shedding it later in the day. By then the zoo would be packed with weekend visitors, but for now the grounds were closed; as serene as the busy staff and a slew of hungry animals could be.

Roger drove along the winding pathways, stopping frequently to empty trash bins into the back of his truck. Because of weekend crowds, Sundays and Mondays were the heaviest garbage days, and every can had to be emptied and replaced before the gates opened. He motored through the Children's Zoo and the Pacific Northwest exhibit. He called good morning to Caesar as he pulled the trash bins from the Feline House and the lion, unimpressed, exposed four-inch canine teeth in a bored yawn. Roger made brief stops at the bear and primate exhibits, took a side trip into Alaska Tundra, and finally turned toward the elephant barn, purposely making it his final destination so he could tarry, if only for a moment.

Parking near the entrance, he hurried inside and grabbed the first bin, located just inside the door. Movement in the exhibit room caught his eye. He turned to offer a cheery "Good morning, campers!" to whichever elephant was inside and gaped in surprise.

Damp with amniotic fluid, an elephant calf blinked myopically, its rubbery trunk hanging flaccid, moving in time to the unsteady sway of its body. Rosy, who'd served as Effie's midwife, rumbled as if amused by Roger's befuddlement. Energized by the sound, the calf tottered drunkenly around its mother's legs.

Roger abandoned the garbage and hurried to find Tucker, who'd just arrived. The senior keeper grinned broadly at the good news, clapped him on the shoulder, and invited him to assist with the calf's initial check-up.

Carrying an ankus—the tool also known as a bull hook, used for centuries to guide elephants, and offer handlers a measure of protection when necessary—Tucker slid between the bars of the exhibit room and motioned for Roger to join him. The cows shuffled closer, drawn by the presence of their senior keeper and obviously intrigued by Roger, a man they'd only seen and smelled from a distance. Gently pushing aside an inquisitive trunk, Tucker ordered them to move back. He greeted Rosy first, as matriarch, and then Effie. Placing a gentle hand on the newborn's wooly head, he asked Roger to hold the calf still while he examined it.

Unable to believe his luck, Roger bent and placed his hands on either side of the elephant calf's ribcage. For the first time, he felt the wiry pelt of hair, the heat of its body, and the thrum of its mighty heart. His eyes stung with tears, and his throat closed with emotion.

Running his hand over the calf, Tucker determined that the little female was about an hour old. Something in his demeanor put Roger on the alert. As a father, he'd grown attuned to such things. He asked if anything was wrong.

Tucker avoided the question. Instead, he gestured at the mess of amniotic fluid and blood on the floor and asked Roger to clean it up before one of the elephants slipped and broke a leg.

Afterward, Roger lingered outside the cage as long as he dared before reluctantly continuing the garbage run. For the rest of that day he ducked inside the elephant barn whenever his duties took him near, just so he could check on "his" elephant. The bad news wasn't long in coming. For reasons unknown, the calf's lower jaw and neck were malformed, making it impossible to nurse from her mother. Christened "Droopy," she was bottle-fed by nursery staff and veterinarian Matt Maberry, but died five days later.

The elephants seemed to accept the loss of the calf better than their human caretakers. Even Effie appeared unmoved, and Roger wondered how much she'd sensed right from the start. He felt sad Droopy had died so young—as the first elephant he ever touched, she held a special place in his heart—but he was grateful she hadn't lingered, captive to a debilitating or painful illness.

That winter brought unseasonable cold. Temperatures hovered around zero and snow fell to a depth of more than fifteen inches. Glacial wind tore through the Columbia River Gorge, coating everything in a relentless patina of black ice, breaking utility wires and tree limbs. The conditions made it dangerous to walk, let alone drive. Roger scattered sand along pathways and in cages, hoping to keep staff and animals on their feet. He shoveled paddocks, chipped ice out of water buckets, added extra straw bedding to cages, and employed a blow torch to open stubborn locks. On days working maintenance, he helped nurse the elderly boilers that struggled to keep interior areas at a livable temperature.

In December 1968, Roger reached the end of his probationary status and was promoted to a permanent employee, which made the future less worrisome to contemplate—particularly after RoseMerrie announced she was pregnant with their second child. Roger continued to shift between departments, but now the assignments lasted longer. He was cleared to relieve senior keepers on their days off, which brought increased responsibility and autonomy. Decisions were his to make without having someone always looking over his shoulder, and any fallout from those choices became his to assume as lead man. The garbage run, manure pile, and hay fork remained a large part of his repertoire, but he didn't mind. As he repeatedly told disbelieving family and friends, "I'm having so much fun, it ought to be illegal."

More time with the elephants gave him greater opportunity to observe and ask questions of Tucker, Robbins, and Cochran. He soaked up every detail, but most impressive was his ability to spout it back verbatim when questioned. He wasn't just curious, he was *learning*. Precious few books about elephants existed in those days, but he sought them out and devoured every one. He developed a solid working knowledge of elephants in general and the Oregon animals in particular, including hierarchy within the herd and who liked what. He discovered that elephants tend to be left- or right-side oriented, as evidenced by which side they prefer to sleep on; they snore; and they routinely expel an eye-watering amount of methane. A favorite joke among the keepers was this: What's the difference between a cheap tavern and an elephant fart? One is a bar room and the other is a *ba-ROOM*!

As his presence in the barn grew, the elephants now greeted him when he arrived. Nothing delighted Roger more than to be welcomed with a chorus of hearty squeals.

Summer 1969 marked his one-year anniversary at the zoo, a milestone Roger celebrated with quiet satisfaction. He'd yet to be permanently assigned somewhere, but that was fine by him because it left open the possibility he might end up where he most wanted to be: with the elephants.

Roger was so wrapped up in his newfound love for elephants, he hadn't given serious consideration to their potential for havoc. He'd never witnessed aggression from any of the elephant cows toward their keepers, and the occasional sparring matches between Belle and Rosy, or Me-Tu and Hanako, were easily quelled by a single sharp "Hah!" from Tucker if they appeared about to escalate. But accidents happened, and keepers could get caught in the crossfire.

Barely a month after Roger's one-year anniversary at the zoo, Belle unintentionally broke Denslow Robbins's arm while attempting to discipline Packy. She'd gone to shove him with the side of her head, missed, and hit Robbins instead. As soon as Belle realized what happened, she squeaked her apologies to the injured keeper. Everyone knew it was an accident, but the event stuck with Roger and left him thoughtful. Belle may have been his sweet, tractable girl, but she could still pose a danger.

Thonglaw, however, was an obvious threat. The bull's hatred of human beings prevented keepers from interacting with him as they did the cows. Roger didn't know why Thonglaw's attitude toward people was so prickly, but he'd been cautioned to keep his distance and did so. While the cows enjoyed regular physical and social interaction with their keepers, Thonglaw remained isolated, unapproachable except by former owner Morgan Berry, who showed up at regular intervals to care for the bull, run him through a litany of commands, and even ride him to prove how well-broken he was. Given the opportunity, Roger wouldn't have climbed aboard for all the money in the world.

***

Later that summer, Roger stood outside a wire-fenced paddock with a group of fellow keepers, looking in at a four-cow herd of Roosevelt elk. Dark brown legs and necks stood in sharp contrast to the creamy beige of their bodies as they nosed through their morning fodder. Despite their size—the largest topped six hundred pounds—these were gentle creatures inured to life at the zoo, displaying interest in their keepers only at meal times.

The muscular bull standing at attention in their midst was another matter. Five feet tall at the shoulder and over eight feet in length, he weighed between 800 and 1,100 pounds. A full forty pounds of that weight crowned his head as a rack of magnificent antlers with a single front-facing prong angled forward like a spear, terminating in a lethal point.

Several weeks earlier, he'd been allowed in to mate with the cows, after which he'd been removed to a separate enclosure.

Yesterday, a keeper mistakenly returned him to the cows' yard, making it impossible for anyone to enter and clear away the night's accumulation of manure. The bull glared at the men and raked the ground with one hoof, silently daring them to try. Torn skin hung from his antlers like shreds of gory cloth, dripping blood onto his shoulders.

The bull was in velvet, an annual stage in the growth of new antlers when the bone is cloaked by a hairy covering rich in blood vessels and sensitive nerve endings. Left alone, the skin eventually dries and sloughs off, or is rubbed away against trees to reveal the mineralized structure of the fully-grown new bone. During the night, this Roosevelt bull caught his rack on something and shredded the blood-rich covering. Senior keeper Lloyd Floessler wanted him removed from the paddock so the vet could treat him with antibiotics and analgesics, and preferred it be done under the animal's own power rather than sedate him. On a normal day, this wouldn't have been an issue, but pain and the smell of blood brought out the elk's aggressive side. No one in their right mind would set foot inside the fence.

Someone suggested a food incentive, but there was already so much fodder littering the ground, the elk would find nothing tempting. After thinking a moment, Floessler suggested someone stand at the window of the feed room and taunt the bull until he came down the chute far enough so the gate could be dropped behind him. The job required finesse because the last thing the man acting as bait wanted was for the bull to become so angered he broke through the four-foot-square opening where his tormentor was stationed, a feat the animal was certainly capable of performing. Meanwhile, other keepers armed with pressure hoses would stand along the paddock's perimeter, ready to blast the elk if the bait wound up in serious trouble.

The keepers' eyes went from Floessler, to the bloody bull, to each other. No one said a word. Finally, Roger muttered something colorful and volunteered. As he saw it, this wasn't a question of fear versus bravery. The job needed to be done, and the sooner it was accomplished, the better.

Taking position inside the feed room, he gripped the rope feeding through the pulley system used to raise and lower the chute door. He'd need to time the drop precisely when two-thirds or more of the bull was inside the chute, otherwise the animal would spook as the door fell and break free. If that happened, they'd be forced to begin again, but with an elk that understood what they were up to.

Nearly an hour elapsed as Roger teased the bull, making runs at the window and then retreating, trying to lure him into giving chase. Light-footed as a swordsman, the animal danced back and forth, snorting and spraying blood, making a proud display of his prowess, but never entering the chute.

Roger grew sweaty and tired. "Come on, you bastard," he muttered. And just like that, as if his words were a magic spell, the pain-addled elk also wearied of the game. Huffing breath, muzzle extended and damp black nostrils working, he moved closer to the chute. One cautious step followed another. His nose passed the entry, then his head and back-swept antlers, and now the bloodstained roll of his shoulders. Roger tensed, breath tight with concentration, and waited for the precise instant when he could drop the gate.

Something grabbed the back of his thigh. Roger yelled, startled by the assault. The bull bellowed, shat in terror, and exploded backward out of the chute. Somehow still clutching the rope, Roger whipped around to see what had attacked him. A trim, bespectacled stranger in a business suit and shiny shoes stood there, grinning with puckish amusement. "Boy, oh boy! When you concentrate, you *concentrate*."

Roger flushed, hot with anger, his hands white-knuckled around the rope. "I ought to kick your ass!"

That infuriating grin never wavered. "Where's your sense of humor, son?"

"God damn it, this is serious business!"

The man nodded, maddeningly unperturbed by the ruckus he'd caused. "You're absolutely right. I'll leave you to it." Cheerily, he slid his hands into his pants pockets and calmly strolled away. Roger watched him go, utterly confounded. When there was enough distance between them to ensure there wouldn't be a replay of the stupid prank, he resumed teasing the now much-wiser bull.

At long last, the animal was contained and Roger stepped outside for a cigarette. Dropping onto a bench, he watched the placid elk cows nose through their breakfast, unperturbed by the morning's theatrics.

A veteran keeper joined him. Flicking a lighter and applying it to his own cigarette, he asked, "You know who you cussed out today?"

Roger jetted smoke toward the sky. "No, and I don't much care."

The keeper grinned and flicked away ash. "You should. That was Jack Marks, director of the zoo."

Roger stubbed out the cigarette, his pleasure destroyed. After everything RoseMerrie had done, all the encouragement she'd provided and the sacrifices she'd been willing to make, if he went home and told her he lost this job because of his damn fool big mouth, there wouldn't be enough left of him to sop up with a towel.

He worked through the rest of the morning on tenterhooks, expecting at any moment to be summoned to the front office. Fate smirked at him when he crossed paths with Marks later that afternoon. Roger averted his gaze, hoping to avoid a confrontation, but the director paused on the path and asked, "You still got your heart set on kicking my ass, son?"

Roger stopped and turned to face him. "No, sir, Mr. Marks, and I'd like to apologize for earlier. I sure am sorry for shooting off my mouth."

Marks smiled. "Well, I may have had something to do with that." He accepted the apology and nothing more was said. Over the years, he and Roger grew to respect and like each other—and Roger came to better understand Marks' spontaneous, sometimes childish sense of humor.

Elephants, giraffes, elk, and now zoo directors—one by one, Roger took them on and learned the lessons they had to teach. At home, he divided his time between rambunctious two-year-old Michelle, his usual duties around the house and yard, and RoseMerrie, whose hugely swollen belly gave the word "pregnancy" a whole new definition. At least once a week, the extended family came together for a meal and laughter.

Summer 1969 rolled on. Tuy Hoa, Rosy, and Pet were confirmed pregnant again, with due dates the following spring. In early August, RoseMerrie delivered a baby girl they named Melissa. Roger handed out cigars and accepted congratulations, making sure everyone understood that RoseMerrie had done all the work.

A few weeks later, he drove up to the barn with a truckload of castoff grocery produce meant for the elephants and walked inside to ask Tucker where he wanted it dumped. Instead of offering their usual happy greeting of warbles and squeals, the cows watched him with reproachful eyes and remained eerily silent.

Worried, he continued through the barn. As he approached Thonglaw's enclosure, the smell of feces and urine overwhelmed him. Inside the cage, the drains overflowed with excrement. The bull stood with his rump pressed into a far corner, head low, looking exhausted and furious.

Roger found Tucker in the office. "What's wrong with Thonglaw?" he asked, the truck of produce forgotten.

Tucker looked grim. He explained that for the past few weeks, Thonglaw had been in something called *musth*, a word of Persian origin that meant "intoxicated." Although not a rut state in the usual sense of a seasonal mating imperative, musth is characterized by a surge in hormones that renders male elephants temporarily insane, brutally aggressive, impervious to pain, reckless, and wholly unpredictable. Episodes of musth begin in the bulls' twenties and occur once a year depending on their age and overall health. Their temporal glands swell and excrete an oily substance with an acrid, penetrating stench that pours down their cheeks and into their mouth. Their penis becomes engorged and dribbles urine constantly, turning the sheath a greenish-white, staining the back legs, and irritating the inside of the thighs. Potentially lethal to anything crossing their path, musth bulls have attacked automobiles, hay stacks, boulders, and even unoffending females of their own species. Episodes typically last ninety days, although they can last much longer.

Bulls in their early teens go through a weaker variety of the condition called *moda*, a word referring to the scent of their temporal gland secretions which, unlike those of fully adult males, smells like flowers. This honeyed aroma broadcasts their immaturity and unwillingness to fight and keeps them out of trouble with older, more combative bulls. As the younger males mature, their odor takes on a sublime aroma that's been compared to a mix of skunk and clover, and they assume the full explosive capacity of musth. It's theorized the true purpose of the condition may be to free the bulls of inhibitions imposed by prior experience, thereby allowing them to challenge the established dominance hierarchy. In captive animals, that inclination to defy the status quo may manifest in confrontations with doors, walls, and keepers.

Common belief in those days held that it was necessary to subdue a captive musth bull by "breaking" him in order to re-establish control. At Jack Marks's invitation, Morgan Berry returned regularly to break Thonglaw. First, he would spray the bull with water. Then, standing safely out of reach, Berry would use a specially-made ankus plugged into one hundred volts to repeatedly "light up" Thonglaw, an operation that could take several hours until the elephant finally stopped fighting and signaled his submission.

Hearing this for the first time in Tucker's office, Roger was horrified. "Isn't there anything we can do to stop it?"

Tucker shrugged. He'd dealt with the situation for years, powerless to intervene. In his opinion, musth was a natural part of a bull elephant's life and should not be tampered with, and he hated seeing Thonglaw in pain. "It's not my decision to make."

The statement confounded Roger. He'd assumed Tucker was the man in charge. Apparently that wasn't the case, at least where Thonglaw was concerned. He didn't yet understand what Tucker had come to realize in his fifteen years at the zoo: namely, that sometimes a man's hands are tied no matter what his title or intentions.

Tucker had learned to live with the things he couldn't change and focus instead on what he could achieve. Increasingly, those thoughts revolved around the end of his career. At sixty-seven years of age, he'd held off mandatory retirement for almost two years by asserting there was no one qualified to assume care of the elephants, but he could buck the system only for so long. Cochran and Robbins were fine keepers, but they would be cruising toward retirement themselves in a few years. Next in seniority was Dale Brooks over in Primates, but he'd turned down the position, having seen an elephant at a different zoo kill a man.

Portland's herd was generally well-behaved, but that didn't mean they couldn't also be dangerous. The most complacent elephant can become deadly given the wrong set of circumstances, and a frightened keeper has little hope of containing a difficult, perhaps life-threatening, situation. Those who fear an animal or are intimidated by it may engage in retaliatory abuse to prove—to themselves, at least—that they're the ones in control.

Because elephants respond to routine and familiarity, Tucker wanted his replacement to be someone committed to staying on the job for years, undaunted by hard labor or long hours. They must be able to manage a crew and guide them into the future without a lot of turnover. They had to speak up on behalf of the elephants even when that meant going toe-to-toe with management. They needed to accept defeat if necessary, and keep working toward a better life for the elephants. They must possess empathy, patience, compassion, and be someone the elephants could trust to not abandon them when things became difficult.

Tucker took little notice of Roger when he first arrived at the zoo. The young father was nobody special, just another newbie. Had Roger left within a few months, fleeing the sometimes unpleasant conditions at the heart of zoo work, Tucker would not have been surprised nor mourned his departure. Instead, Roger advanced in his duties and responsibilities at an impressive rate without losing any of his enthusiasm. And it was obvious he adored the elephants.

On his next rotation through the barn, Tucker called Roger into the office and laid it out plain and simple: retirement called and he wanted Roger to replace him as senior man in the elephant barn.

Roger's mouth went dry with desire. "How do Joe and Denny feel about this? I don't want to step on any toes."

Tucker assured him his fellow keepers thought it was a great idea.

"Then, yes, sir," Roger said, awash in shock and feeling unaccountably blessed. "I'd like that very much. Thank you."

Tucker waved his gratitude aside. "Don't thank me; you earned it." He picked up the telephone and called Bill Scott to announce his decision. He asked for an additional year in which to train his replacement and was granted an extension on his contract.

That night, the entire Henneous clan erupted into celebration. Beaming with pride, RoseMerrie called every relative and friend they had to share in the good news. She'd never doubted Roger for an instant.

Chapter *Five*

ELEPHANT 101

1970

Once Tucker decided that Roger would be his replacement in the barn, he wasted no time. They had one year to train. No more extensions would be granted. Those in charge made certain he understood that.

When Roger arrived the day after accepting the promotion, the old man was waiting for him. Tucker told him to grab an ankus and disappeared down the hallway.

Roger ducked into the office, snatched an ankus off the wall, and caught up with Tucker outside the front exhibit room where four of the barn's eight cows contentedly searched through scraps of hay for any kernels of succulent grain leftover from breakfast.

Belle and Effie looked up as the men appeared, dismissed them in an instant, and returned to foraging. Me-Tu rubbed lazily against the bars, scratching an elusive itch. Pregnant Pet shuffled over to stand directly in front of the keepers, raised her trunk to her forehead, and opened her mouth, exposing a wet swatch of pink tongue and molars roughly as long and wide as Roger's hand. Elephants are born with six sets of teeth embedded in their gums like a row of train cars. One set at a time is used to grind their food, breaking down the tough outer layers of vegetation. As the old teeth wear down, they drop out and the next set moves into position. In the wild, any elephant that doesn't die of injury, illness, or human depredation will eventually starve when the last set of teeth becomes useless and they can no longer feed.

Motioning for Roger to join him, Tucker slid between the bars of the exhibit room. In seconds, they were surrounded. The cows gazed down with gentle curiosity, rocking easily from side to side like a field of wheat stirred by a breeze. Everywhere Roger looked, he saw gray. He was suddenly, deeply aware how one wrong move might turn him into a smear on the floor.

This close, it was easy to see the elephants were not uniform in color. Their skin ranged in shade from the dark gray of a storm cloud to a soft pinky-beige. The edges of their ears, eyes, and across the base of their trunks were particularly pale, and dotted with black freckles. Their broad heads cocked from side to side as they focused first one eye and then the other on the men. The cows rumbled, squealed, and trilled, conversing softly among themselves, remarking on this event. Their muscular trunks—which serve as nose, upper lip, and hands—coiled and twisted, snuffling boot soles, uniform pants, faces, and hair, collecting organic clues on what the men last ate, their physical condition, even their emotional health. As much as two-thirds of an elephant's brain is devoted to olfactory intelligence, matching scents with memories of past experiences, and they use that information constantly.

It took every ounce of fortitude Roger possessed to remain motionless and silent during their inspection, particularly when those impertinent nostrils suctioned his armpits and groin. Speaking in 2016, Roger said, "Elephants have no manners whatsoever, and that trunk will go where it pleases. But, boy, when they're checking out your gender, it's a bit disconcerting."

Roger felt an odd vibration in his chest and massaged his sternum. Tucker told him to place his hand on Belle's forehead, right between the dips above her eyes. When he did, the sensation intensified. Roger's eyes went round with wonder. He'd read about this phenomenon in books written by "Elephant Bill" Williams, but now, experiencing it for the first time, he finally understood.

This was infrasound, the silent speech of elephants; a wave frequency too low for humans to hear, but which can be perceived by sensitive individuals. An elephant can adjust the volume and rate of this communication by opening and closing their mouth, moving their ears, changing the position of their body, and raising or lowering their head. Infrasound can travel for up to twenty miles and pass through trees, rocks, and buildings, enabling elephants to communicate even when they can't see one another.

Tucker asked Belle to back up, which she did immediately, and the men slipped free of the scrutinizing circle. As they passed from the exhibit room into the keeper alley, Tucker flashed a brief smile. "If you're going to work with them, they need to know you," he said. "Now it begins."

Because of Belle's equable disposition, he moved her to a separate enclosure to teach Roger how to correctly use the ankus. At the Oregon Zoo, these were cane hooks or axe handles outfitted with a stylized hook mounted on one end. Tucker explained that the tool was meant to guide the elephant, not serve as a weapon to torment or brutalize. Using it involved reaching beneath the trunk, or jaw, or behind an ear and putting on a bit of pressure with the hooked end—nothing extreme, just enough to get the elephant's attention and communicate that you really need it to work with you.

He stressed gentleness and taught by example. An elephant's skin may appear rough and dry in texture, but it's surprisingly soft and supple, and feels somewhat like an eraser. At the rear end, their hide is about three centimeters thick, but around the mouth, anus, and ears it's much thinner and can be easily torn.

Tucker guided Roger's hand to Belle's mammary glands, armpits, and the back of her ear where the skin felt like silk. Roger nodded. If he wasn't careful, he could seriously hurt or injure one of the elephants. He swore to always be careful, and set about learning the primary Laws of Tucker:

- More can be achieved in this world with kindness than with brutality.
- Don't try to out-muscle them because you'll lose. Out-think them instead. Offer them a better deal.
- Be fair because elephants understand fair.
- Maintaining control is an exercise in intellect.
- Abuse is the lazy man's solution to a problem.

When difficulties with the animals arose, Tucker and his keepers worked hard to cross the divide between species and see the problem from the elephants' perspective. They cajoled, wheedled, and offered succulent incentives such as melons and bamboo. When an elephant needed to be introduced to something strange or upsetting, the elephant's preferred keeper became the one to do it. Every moment of every day was based on trust. Tucker wasn't above yelling, or whacking a haunch, shoulder, or leg with the handle of his bull hook, but that action was reserved for moments that could turn dangerous for the elephant or its handler. Indiscriminately "laying on the lumber" earned a keeper nothing. The trick was to begin with small corrections and a lot of sweet-talk. If an elephant was being particularly stubborn or turned dangerous, you left them alone, assuming you could do so safely, and gave them time to consider their options. A keeper could always up the ante if necessary, but any room for negotiation vanished if one began at the extreme.

"Nobody can force an animal that big, that strong, and that smart to do something it's opposed to, not without things getting ugly," said Roger in 2015, when asked about the ethics of using the ankus. "On the other hand, you can't allow the elephant to run the show. Tools like the ankus give humans an edge when necessary, but the less it's used, the better for both elephant and keeper because not using it forces you to work harder at communicating."

In those days, there were no schools where people learned how to work with elephants; training was on-the-job. Too often, inexperienced men and women were handed an ankus and told to get to work without any genuine understanding of what that meant. Some of those handlers were tough, others fearful, and many never learned to put thought before action. Instead, they let the ankus do the talking and consequently, some elephants were brutalized.

Tucker didn't believe in abuse, which to him meant repeated hard blows or intentionally causing pain, and he taught his men the same. "I never cared for the bull hooks myself," Roger added. "I got chewed out more than once for not having one handy." Over time, his favorite attention-getting device became a simple church key—a can opener with a pointed end used to pierce the metal tops in those days before pop tabs—that he wore on ring hanging off a belt loop. By applying the pointed tip with just enough pressure to exert a little pinch, but never break the skin, Roger could gain the attention of a fractious animal and bring it back on task.

Another Tuckerism was this: Pay attention, because elephants are subtle. They could have an off-day same as any human, and all manner of things might cause them to misbehave, including boredom, physical discomfort, a shift in herd dynamics, or a change in keepers. Even playfulness, if not carefully regulated, could become a cause for concern.

Elephants possess a refined sense of humor. They love to play and tease not only among themselves, but also with their keepers. Some were particularly adept at stealing tools from a distracted handler in order to initiate a game of keep-away. But even something as seemingly innocuous as a trunk swat could have real challenge behind it, considering the trunk of an adult elephant can weigh as much as 300 pounds.

Pet was a perfect example. Whenever the barn crew gained a new member, she would behave beautifully for about six weeks, long enough to instill confidence in the rookie or lull them into a sense of complacency. Then, without warning, she'd thump them with her trunk hard enough to make them stagger.

"If you wimped out and backed away, she owned you," Roger recalled. "You'd never get her to do anything after that because you'd given her control and she wasn't about to give it back. But if you got in her face and roared, really shouted, she'd drop her ears and *eek-eek-eek* and from then on she was your best friend."

Roger welcomed the challenge Tucker's philosophy presented because more than anything, he wanted that special connection the older keeper had with the elephants, a bond which enabled him to sense issues before they became a real problem. Constant vigilance was key, and Tucker's eyes were never still.

"It's like his head was on a swivel," Roger reminisced. "Al could read the elephants' body language and understand what was about happen almost before they did." Often a sharp "Ack!" barked from the back of his throat was all it took to restore calm.

Roger had yet to master that powerful talent the day he was sent to shut off the water feeding into the cement wading pool in Thonglaw's yard. The procedure, simple enough at face value, involved climbing down a vertical steel ladder into a concrete bunker, reaching into a recess about a foot off the floor, and turning the wheel valve. It was obvious someone inexperienced in the ways of elephants had designed the system, because the bunker was located in the yard adjacent to the pool, where an intelligent and determined elephant could extend its trunk and harass whoever was down there. The optimal solution—to relocate the bull to another area whenever the water flow had to be adjusted—wasn't practical because that might need to happen several times a day, and Thonglaw rarely agreed to shift without argument.

Roger warily kept an eye on the bull as he prepared to descend the ladder. Belly-deep at the far end of the pool, Thonglaw idly sucked up water with his trunk and leisurely sprayed it across his back. "He was a remarkable animal," Roger recalled. "Beautifully proportioned and absolutely gorgeous," but dangerous as hell.

Thonglaw cooling off in his pool, circa 1960s. (Personal photo)

As he swung over the lip of the bunker, Roger gave the bull a final look and rapidly descended. The instant his boots hit concrete, however, a wave of cold water poured out of the pool and into the bunker, drenching him. Thonglaw was on the move.

Roger shrank against the far wall as the bull's trunk lashed into the opening, nostrils working feverishly as he sought to lay hold of the keeper. As more water cascaded over the rim, Roger crouched and backed into the wheel housing as far as he could in an effort to present a smaller target. A third wave of mineral-smelling water gushed into the hole. Roger swiped it from his eyes with one hand and with the other slid the buck knife on his belt out of its sheath. The last thing he wanted was to injure Thonglaw, but no way was the elephant going to take him without a fight.

Each time the trunk came close, he jabbed it with the blade. The bull didn't even seem to notice. "Riding out monster storms in the North Atlantic never particularly frightened me," Roger recalled, speaking about the incident during an interview in 2016. "But that day in the bunker, I've never been more terrified." Roger somehow kept his head, almost as if a part of him had stepped aside to observe as he fought for his life, and he soon realized something very important: While Thonglaw could smell him and hear him, the proximity of the bunker to the pool and the angle at which it was constructed prevented the bull from seeing Roger. If he could squeeze a bit further into the recess and keep the trunk from latching onto him, surely someone would come along sooner or later to rescue him.

Ten minutes ticked by, then twenty. Roger's bent knees, back, and ankles grew stiff and painful. Each drenching wave temporarily blinded him, plastering his uniform against his body. His knife-wielding arm grew exhausted, heavy and leaden, and his fingers cramped around the hilt. He shivered from terror and the bunker's dank chill. When he wasn't swearing at the bull, he yelled for help, voice echoing in the small chamber. Through it all, Thonglaw remained eerily silent, the only sound the whoosh of air being sucked up his trunk with every breath.

Back in the barn, Tucker wondered where Roger had disappeared to. It wasn't like him to wander off or shirk duty. When the senior keeper suddenly remembered where he'd sent his trainee, he swore, snatched up a pitchfork, and raced to the exercise yard bellowing, "Thonglaw, back!" even before he saw where the bull stood. Down in the hole, Roger nearly wept with relief.

Prone as he was to challenging authority, this time Thonglaw decided against it. He abandoned his game and withdrew to the far end of the pool where he watched with sullen resentment as Roger shakily emerged from the bunker.

"Did that big bastard hurt you?" Tucker asked. He glanced at Roger, trying to ascertain any damage while keeping one eye on the bull at all times.

Roger shook his head. "Only my pride."

If Tucker worried the experience would send his recruit running in the opposite direction, he couldn't have been more wrong. Roger was too stubborn, and his desire to work with the elephants too great, to call it quits. As he toweled off in the office and poured a revitalizing cup of coffee, Robbins and Cochran convened for a round of good-natured ribbing. They'd survived encounters with Thonglaw and were relieved nothing worse happened to Roger than an impromptu shower and a hair-graying dose of fear. If he hadn't been one of them before, he surely was now.

Roger swallowed a scalding mouthful of beverage, relishing the burst of heat in his belly, and asked what they were going to do about the cuts he inflicted on Thonglaw's trunk.

"Nothing," said Tucker. They'd keep watch to make sure the wounds didn't fester, but otherwise those cuts would be left to heal on their own. Any discomfort would serve to remind the bull Roger was no pushover.

Unfortunately, Thonglaw wasn't the only problem elephant in the barn. Effie was proving to be a burden as well. The original agreement between the Oakland and Portland zoos had been for the cow to be bred to Thonglaw, produce a calf, and return to California once construction on a new $40,000 elephant compound was completed. After the death of her first calf, Droopy, Effie failed to conceive a second time, and the City of Oakland never put up the money for the improved enclosure. When the Oregon Zoo requested to send her home, the Oakland Park Commission refused, having enjoyed two years free of the cantankerous cow. Instead, they voted to get rid of her at the least possible expense.

The Oregon Zoo didn't want her. They already housed nine elephants of their own, three of which were pregnant and due to deliver in the next few months. Twelve animals would strain the capacity of the eleven-year-old barn. Effie needed to go. Busch Gardens expressed interest, so in February 1970 she was loaded onto a truck bound for Florida.

The birth of the calves, one after another, brought both joy and heartache. Tuy Hoa delivered a 310 pound stillborn daughter in March 1970. (A necropsy revealed a partially collapsed right lung and a spherical mass inside the left lung.) April 1970 brought Rosy's daughter Tina, and Pet's daughter Judy arrived in May.

Roger found little difference between waiting for his wife to deliver a baby and waiting for an elephant to deliver a calf. Both involved lots of coffee, lots of cigarettes, lots of worry, and little sleep. Until then, his only experience with elephant pregnancy was his single brief encounter with Droopy. Now he was present when labor began and stayed through delivery no matter how long it took. He measured the calves and weighed them, watched them take their first steps and swallow their first mouthful of mother's milk.

While the brains of most newborn mammals weigh approximately 90 percent of the adult brain of their species, in elephants that amount is reduced to around 35 percent. Considerable growth must take place to achieve the ten-to-twelve-pound brain of an adult, and during this time, a calf is dependent upon its mother for everything. It takes about six months before a newborn calf can coordinate its movements, find its mouth with the tip of its trunk, and eat and drink without slopping its food and water. Incapable of feeding solely on solid food until well into its second year, it relies on its mother's milk to make up the bulk of its diet. On this alone, a calf will gain between twenty-two and forty-four pounds of body weight per month.

Elephant milk appears thin, but it contains more sugar and less water and butterfat than the milk of a dairy cow. In fact, milk from dairy cows should never be used as a substitute for orphaned elephant calves because they're unable to metabolize the high butterfat content. The dairy milk congeals in the calf's stomach as a pasty, semi-solid mass, and the calf literally starves to death while being fed. Formulas now exist to help keep orphaned elephant calves alive until they can begin to feed fully like an adult.

Whenever he had a moment free, Roger paused to watch Tina and Judy at play. Elephant youngsters enjoy a good romp and will devise games using whatever they find lying around. They delight in tomfoolery and mischief, and happily torment tolerant adults or older calves. If no pachyderm playmate is available, they'll entice the nearest handler into joining a game of keep-away. Ears extended, trunks and tails gyrating, the calves darted around their enclosure, squealing in delight.

As the spring of 1970 ended and summer approached, Roger focused on his training and did his utmost to ignore the growing unrest infecting Portland. A massive weeklong antiwar demonstration held by Portland State students at the South Park Blocks erupted into bloody violence and race relations issues flared, threatening to burst into flame. A newspaper article called the downtown area a bomb-site, but up at the zoo, everything remained peaceful.

That peace was shattered on July 4 when a drunken Independence Day celebration prompted nineteen-year-old Roger Dean Adams and two friends to climb the locked gates of the zoo and go exploring. After causing disruption in several exhibits—including hanging from the ledge of the grizzly bear grotto and throwing a small penguin into a pool—Adams dangled by his hands above the lion pit. At the bottom of the sixteen-foot drop, Caesar and his mate Sis prowled with heads lifted, watching and scenting the young man. As Adams swung, legs kicking, taunting the lions with the enticing length of his body, Sis crouched, hindquarters tense and quivering, then leapt, sank her teeth and claws into Adams, and dragged him screaming into the pit. By the time one of his friends alerted the security guard, Adams was dead. The following morning, zoo director Jack Marks told reporters anything he cared to say would likely be unprintable. After all, he said, the lions were where they belonged.

Two days later, Lorin Floessler arrived for work and found Sis and Caesar bleeding from multiple gunshot wounds. Despite swift veterinary care, both of his beloved lions perished, leaving him heartbroken. It was one thing for an animal to die from illness or advanced age, and quite another to lose it to a violent, retaliatory act.

The event set the zoo community and the city reeling. It wasn't until two years later that one of Adams's companions from that awful night walked into a Portland police station and confessed to having shot Sis and Caesar to pay them back for killing his friend. The young man was sentenced to three years' probation, the first year to be served in jail on a work-release program, and ordered to pay the zoo $1,200 in restitution—about $7,800 today.

Everyone at the zoo was hyper-vigilant after the tragedy. Locks were checked more frequently than usual, and the tops of fences were reinforced to dissuade climbers. The thought that the elephants might be at risk, and the knowledge of what could've happened had the trespassers gained access to the barn, kept Roger awake for several nights, prompting RoseMerrie to pointedly suggest that he go sleep at the zoo rather than keep her awake with his tossing.

As his year of training progressed, Roger discovered there was more to the role of senior elephant keeper than just keeping elephants. Not only was there more paperwork than he liked, he was also expected to interact with zoo visitors and answer their questions. He didn't consider himself much of a front man, but he was gregarious and often funny, and people seemed drawn to him. Roger understood that good community rapport translated into increased attendance, which meant more money for the city and, he hoped, a larger budget to provide for the animals. Because there were always children in the audience, he made sure to talk slowly and choose his words with care.

There were other extraneous duties, such as school visits, fundraisers, and sitting on committees, none of which Roger particularly enjoyed, but the one job he looked forward to was working with Packy and gaining his trust. Until now, that duty had been Tucker's. He'd taught the young bull the basic cues every elephant learned to help facilitate their daily care: lift feet, lay down, back up, sit, and turn around. He'd also trained Packy to perform various tricks like front and rear leg stands, head stands, and walking the plank, which were used to entertain visitors during daily shows in the elephant yard.

"I enjoyed working with Packy, but I hated those performances," Roger said, thinking back to those early days. "It's degrading to make elephants perform, and the white costumes we had to wear were ridiculous. But back then, every zoo, carnival, roadside show, and circus had an elephant act, so we did, too. Visitors not only expected it, they loved it."

As the new senior keeper, Roger would take over the duty of performing with Packy, but before that happened there was a lot of work to do. Packy knew the routine cold and probably could have given the show solo, but Roger needed training. As it turned out, so did Tucker—in letting go. Relinquishing his central role in Packy's life proved to be more difficult than he imagined. He'd nurtured the calf, played with him, trained him, and seen to his every need. Packy so loved Tucker in return, he once managed to open an outside door with his trunk in an attempt to follow him home at the end of the day. The long-legged eight-year-old bore little resemblance to the endearing wooly wind-up doll Roger remembered, and increasing episodes of stubbornness and pugnacity hinted at further changes to come, but for now things remained status quo.

They practiced every day behind the scenes, Roger and Packy in the yard and Tucker standing in the barn's open doorway, until the old man felt confident in Roger's ability to control the bull. Packy seemed amenable to the change in handlers and responded well to his new partner. It never occurred to either man that Packy's willingness and easy manner might be tied to Tucker's presence.

At long last, the big day came for the new team to make their debut. Beneath a balmy spring sky, the elephants and their handlers paraded into the yard to an enthusiastic round of applause from the audience lining the rail. Belle, who'd performed since infancy and could, as Roger put it in 2017, "do anything but cook breakfast," partnered with Denslow Robbins. Me-Tu teamed with Joe Cochran, and Packy brought up the rear with Roger. Hidden among the audience of several hundred spectators, Tucker watched.

The performance went off without a hitch until Packy realized Tucker was nowhere in sight. He turned from Roger and ran toward the barn, seeking his favorite keeper's familiar, lanky frame. The closed door prevented him from entering, but he pressed an eye to the crack, anxiously searching.

Aware that everyone, especially his boss, was waiting to see what he would do, Roger called, "Packy, come here!"

The elephant ignored him.

"Packy!" he repeated sharply. "Come here!"

The bull whirled to face him, head lifted, trunk curled up, ears fanned. Roger's mouth turned cottony and his hands to ice. Packy was demonstrating classic body language before a charge. The bull never gave any indication he had a problem with Roger, and Tucker never talked about this in their sessions together, so what the hell should he do? His brain whirled with escape options. He could dash to the moat if Packy charged, or dodge behind Belle, but if he did either, that would give the bull carte blanche to do as he liked despite what he was told. They could wind up with dead men on the ground, broken elephants in the moat. Even if everyone survived, Roger would never regain the ground lost with Packy. After all, he couldn't force the young bull's compliance, only encourage it. Worse, Tucker would lose faith in him and maybe decide he'd chosen the wrong man to be his replacement.

Roger's spine stiffened. That was *not* going to happen. He wasn't about to piss away a job he loved just because some upstart, adolescent elephant thought it was okay to throw a temper tantrum. Drawing breath deep into his lungs, he walked forward. "Packy, come!" He was grateful to hear his voice sound more in control than he felt, and hoped the crowd couldn't tell he was nervous.

Packy held his ground, bulging eyes riveted on the approaching keeper.

Beyond the rail, the crowd waited in silence, mesmerized by the battle of wills. Somewhere behind Roger, the two cows stood quietly with their handlers, motionless and watchful as he pressed forward, sternly issuing the command every couple of steps. The space between man and elephant narrowed until the air nearly boiled with emotion. Less than two body lengths separated them. If Packy charged, there'd be no room to avoid him.

"Packy." Roger grated out the name, making it as rough, loud, and angry-sounding as he could without shouting. "You...come...here...*now*."

For a tense moment, the tableau held. Then, as if a pressure cock released, Packy's ears dropped, his stance relaxed, and he squeaked surrender. Roger's legs turned to gelatin. Shaking, he closed the gap between them, hooked his ankus over the base of Packy's trunk, and firmly drew the bull's head down to where he could speak directly into his ear. "Now, goddamnit, you listen to me!" he said through clenched teeth.

Packy's performance was flawless for the remainder of the program. After the three elephants retired to enthusiastic applause, Roger returned his bratty partner to his enclosure and sat down in the office for a much-needed cigarette. It wasn't long before Tucker appeared. "You realize how close that was, boy?" he asked.

Roger blew a lungful of smoke at the ceiling. "Only my laundress will know how close."

Tucker's laughter held a dry edge of familiarity. "When he broke, I was halfway into the moat, ready to climb the wall and snag him, but you did exactly the right thing."

Praise from a man he deeply admired meant the world to Roger. Facing down Packy's challenge had been an important obstacle to overcome, proving to the bull that Roger meant business. He felt confident about performing with Packy again, but it was not to be. Someone on staff witnessed the mutiny and reported it to Jack Marks. Concerned lest a disaster occur in front of an audience as Packy became more aggressive and unpredictable with age, the zoo director declared an end to the bull's show business career.

Roger wasn't the least bit disappointed.

For the remainder of 1970, Tucker labored to distill his mountain of hard-won knowledge and pass it on to Roger, who struggled to imprint it on his bones. Both felt the pressure of time passing and the looming certainty of the old man's retirement.

Computers did not come to the barn until 1998, so keepers updated each elephant's file daily in writing, noting their physical condition, diet, appetite, fluid intake, behavior, and whether blood or other bodily fluids were being drawn for use in research. Working with Tucker, Cochran, and Robbins, Roger learned about bedding, diet, and the intensive labor involved in maintaining the integrity of an elephant's feet, their most vulnerable body part. He was taught to bathe them and care for their skin; memorized which hay vendors to patronize and which to avoid; became adept at cajoling the animals onto a truck scale to be weighed; and began to acquire that special ability to read their body language in order to stay one step ahead in the game.

Too soon, the day arrived when both he and Tucker realized that going forward, Roger's greatest teachers would be the elephants themselves. It was time for Tucker to move aside and let the new senior keeper take over. His final stroll through the barn to say goodbye to his four-footed family and two-legged friends was a somber occasion. For the first time, Roger thought Tucker looked old. He was already in his mid-60s when they first met; thin and gray-haired, but upright and energetic. Now he seemed diminished—stoop-shouldered, slow-moving, and reluctant to leave. Watching him go from cage to cage and elephant to elephant, his big hands stroking the soft skin around their eyes and behind their ears, Roger felt a lump the size of a truck grow in his throat. When Tucker turned to shake his hand and asked him to please take good care of his boys and girls, Roger feared they would both break down and bawl like babies. The barn wouldn't be the same without him.

But old habits die hard. Despite a nice home at the beach, the company of family, and regular attendance at grange dances with his wife, Tucker found it impossible to stay away. He missed the elephants, of course, but what he craved most was the camaraderie of his fellow keepers, their shared stories and experiences. Whenever he visited—far more often than anyone had expected—Tucker did his best to not get underfoot and always offered to lend a hand. And who could refuse him, the man who'd built the elephant program from scratch? Belle, Thonglaw, Packy, and even Morgan Berry were important—there likely wouldn't have been a breeding program without them—but no one could deny Tucker's impact on the elephants he loved and the men he trained. His decision to choose compassion over force made all the difference, and Roger credited him more than anyone else for developing such a contented and well-behaved herd.

Tucker knew he was no longer in charge and behaved like a model visitor in every way but one. Where Packy was concerned, the old man acted as if the good old days had never gone. Without thinking twice about it or asking permission, he'd blithely enter the maturing bull's enclosure to play with him, put him through his old routine, and make him stretch out on the ground for a belly rub.

Packy obviously enjoyed these reunions and never displayed the least amount of aggression toward his former keeper, but the episodes, as Roger recalled in 2016, "put the shit right up my back." He knew he ought to stop them, but found it impossible to give orders to the man who'd made him, the one who'd placed him where he was today. He couldn't deny Tucker what he so obviously enjoyed and maybe even needed. On the other hand, it would be a damn sight more difficult to explain to the governing board if Packy suddenly chose to reduce the old man to a puddle of jelly. Roger stood by during each visit, grinding his teeth in agitation, and walked away when he couldn't bear to watch a moment longer.

All those years with elephants had honed Tucker's intuition to a fine point, so it wasn't long before he picked up on Roger's distress. One afternoon, a few months after retirement, he showed up in the office and apologized. Playing with Packy had been his privilege when he was in charge, but doing so now left the new senior keeper with one hell of a story to invent if something went wrong. He promised never to do it again, and his visits ended shortly thereafter.

Chapter *Six*

HALF A TON OF MANURE AND OTHER STORIES

1971–1973

Roger always believed Tucker's decision to stay away had more to do with astringent self-perception than with Packy. The old man realized that no matter how good his intentions, eventually he couldn't help but get in the way. As work in the barn evolved over time, he was bound to argue about something or offer one opinion too many. Even so, Roger missed his staunch support and gentle guidance, particularly in the early months as the reality of what he'd taken on began to sink in.

When Roger assumed leadership at the beginning of 1971, there were eleven elephants in residence: Rosy, Tuy Hoa, Thonglaw, Belle, Pet, Packy, Me-Tu, Hanako, Winkie, Tina, and Judy. Each drank up to fifty gallons of water per day and received a veritable smorgasbord of nutrition, including one hundred pounds of the best timothy hay, forty pounds of mixed fruit and vegetables, and three pounds of crimped oats laced with vitamin supplements and salt. Bamboo was an occasional treat provided

Roger speaks to visitors alongside the best-behaved elephant in the herd, Belle. (Personal photo)

by Park Bureau gardeners, who cultivated the invasive plant to provide the zoo with jungle ambiance. The elephants eagerly gobbled the leafy fronds, but refused to eat any stalks cut from the feline cages. Humans are elephants' only natural predator, but tigers and lions will follow a lone female ready to give birth in hopes of snatching her infant. Even elephants that have never lived in the wild react to the rank smell of cat musk.

Melons were another favorite treat. When they were in season and cast-off items began to arrive from local grocery stores, Roger would line the cows up along the rail with their trunks against their foreheads, mouths open to receive a watermelon or cantaloupe. One chomp was all it took to reduce the fruit to pulp and spray bystanders with sticky juice. One-year-olds Judy and Tina didn't possess the size or strength to crush the melons with their mouths, so they devised a different method. It didn't take long to discover that kicking the melons, sitting on them, or landing on them in a full-body slam brought succulent results.

Adequate food and water translated into almost a ton and a half of excrement per day, all shoveled and carted away by Roger, Cochran, and Robbins. Adult elephants defecate up to fifteen times a day, resulting in roughly two hundred and fifty pounds of manure per animal courtesy of the inefficiency of their digestive tract. The elephant gut lacks symbiotic bacteria to aid digestion. They can consume up to six hundred pounds of vegetation in a single day, but draw only about 40 percent of the available nutrients, losing the remaining 60 percent as manure. Hence the need to feed them frequently and well. Add to that the daily requirements of repeatedly cleaning the barn, bathing and scrubbing the animals, shifting them indoors and out, and the myriad other tasks needed to keep things running smoothly, and it's no wonder the elephant keepers fell into bed each night exhausted.

Tired as he was, Roger never stopped loving the animals or the work. What he didn't love were some of the changes coming to the zoo. In 1971 Jack Marks signaled his intent to retire after a record-setting and innovative twenty-three years at the helm. His successor, Philip Ogilvie, served as director to the Oklahoma City Zoo and Botanical Garden, as well as the Minnesota Zoological Garden. He certainly had the experience required to step into Marks' shoes, but Roger was more interested in attitude than he was titles and credentials. He laid those feelings bare at their first meeting. "I can work *for* you or I can work *with* you," he told the new director. "It makes no difference to me, but I think I can contribute more by working with you." Roger explained that, as he saw it, his job was to provide the elephants with the best possible care no matter who sat in the big chair. "Management might change, but the zoo goes on."

Denslow Robbins and Joe Cochran announced they would also leave soon, a decision prompted by a shift in the zoo's governing body from City Civil Service to the Portland Zoological Society. This change had dozens of long-term city employees rushing to transfer to other city-run departments out of fear of losing their pensions. As a relatively new hire, Roger was little affected by this shift and would build his retirement through the new governors.

He'd deeply miss the men who'd helped train him, but didn't have time to dwell on thoughts of their departure or who might take their place. For now, he had larger worries.

For all their great size, elephants are vulnerable to a number of health problems including tetanus, tuberculosis, parasites, dental issues, heart disease, pneumonia, colic, and psychological trauma. In captive animals, whose opportunity for exercise is limited by available space, obesity is a constant concern, particularly for those receiving a diet of highly-digestible concentrates rather than the fibrous vegetation offered at the Oregon Zoo.

New health issues have also arisen. Over the past twenty years, Elephant Endotheliotropic Herpesvirus (EEHV), a chronic, usually latent disease present in most Asian elephants, has become an increasing concern in both captive and wild populations. When it becomes active in a juvenile elephant, it can cause devastating hemorrhagic disease that is usually fatal. A report published in *Journal of Wildlife Disease* in 2013 estimates that in Europe and North America, 65 percent of deaths of Asian elephants between the ages of three months and fifteen years have been due to EEHV. Although a blood test to detect the virus was developed in 1999, and zoos routinely send samples to the Smithsonian's National Elephant Herpesvirus Laboratory, once the virus becomes active, there is little time in which to attempt treatment with fluids, transfusions, and antiviral medication. At present, there is no vaccine, and EEHV is the leading cause of death in Asian elephants under the age of eight in North America. In November 2018, the entire Oregon Zoo community mourned when Lily, the youngest member of its herd, succumbed to the sudden onset of the disease one day short of her sixth birthday.

There is hope, however. Although the Indianapolis Zoo lost the two youngest members of its herd to EEHV in March 2019, two other elephants there diagnosed positive for the virus and have been treated successfully.

By far the greatest health challenge Roger faced was how to maintain the integrity and functionality of the elephants' feet. Elephants appear flat-footed, but their heel is actually a pad of fatty and elastic connective tissue. The bones of their feet are positioned so they actually walk on their toes, as though wearing high heels. Foot circumference increases up to four inches when bearing weight, which aids in pumping venous blood from the foot on its return to the heart.

Denied regular exercise, elephants can grow a thick callous-like layer on the bottoms of their feet which hardens over time. This callous must be trimmed regularly to prevent it from wearing down unevenly and rendering the foot and the animal's gait crooked. Other factors contributing to foot issues include nail overgrowth, improper walking surfaces, excessive moisture, insufficient foot grooming, unsanitary conditions, inherited poor foot structure, malnutrition, and skeletal disorders such as arthritis. Free-ranging elephants also encounter foot problems in the form of snare injuries, lacerations, fractures, and penetration by foreign bodies. Elephants in the work camps of India and Southeast Asia experience foot issues similar to those of zoo and circus elephants.

Roger's favorite tools for maintaining foot health included his well-used farrier's rasps and horse hoof nippers, a hand-held grinder fitted with an abrasive pad, hoof knives, and an impressive array of makeshift implements created on the fly as needed. Feet were given a cursory examination daily. A more thorough inspection and treatment occurred weekly, a process that could take up to an hour per foot. First, embedded debris such as dirt, stones, and hay was cleaned away. Split nails were treated with antibiotics, and broken areas filled with bonding material such as acrylic patches or epoxy. Nail overgrowth was rounded in order to properly redistribute pressure across the entire pad of the foot, while abscesses and lesions were treated with oral medication, warm foot soaks, and trimming of any necrotic tissue. Extreme caution was taken not to pare down an area so radically as to change the angle of the foot, which would then cause further complications.

Every elephant in the barn, excepting the calves, suffered foot issues of one sort or another requiring treatment, sometimes on a daily basis. The lone exception was Thonglaw, who refused human contact unless forced into it by Morgan Berry. Sixteen-year-old Tuy Hoa showed early signs of arthritis, but Roger didn't know if this was genetic pre-disposition or whether she'd missed some particular balance of nutrients early in life, an extremely likely scenario given that she was taken from her mother at an early age. Tucker provided for her as best he could, but had only a rough idea of what her growing body required.

Affable Pet presented a different sort of problem. For reasons unknown, she refused to place her feet on a tub, akin to a half-barrel flower box turned upside down, to have them examined and treated. The difficulty predated her arrival at the zoo, so its origin remained a mystery. She wasn't averse to having her feet touched, and even seemed to enjoy it, so Roger suspected the problem lay with her pigeon-toed stance. Perhaps the in-turned structure of her feet made lifting them onto the tub difficult or painful. Placing a foot on the tub allowed the elephant's toenails to project over the edge, where they could be more easily examined and rasped, and Pet's refusal to do so made it harder for keepers to maintain them.

The struggle to win Pet's cooperation left everyone frustrated and unhappy. Roger mulled it over one day while working and hit upon a brilliant idea. The elephant yard was built with three tiers rising like stairs. If someone positioned Pet on a higher level with her toes hanging over the edge, would she allow Roger to trim her feet from below?

The solution worked beautifully. Pet stood calmly with one of the keepers, contentedly consuming every carrot slid into her mouth while Roger worked, and happily cooperated with regular pedicures from then on. One day, however, she jerked—he may have accidentally poked a sore spot—and broke the blade off the Stockman knife he was using. Because it was a treasured gift from his parents, Roger mailed it to the manufacturer with a note of explanation, assuring them the damage was due to simple abuse on his part rather than failure of their product and asking that they please replace the broken blade at his expense.

A few weeks later, a package arrived containing a new knife, compliments of the Stockman Company. Roger was delighted until Pet managed to snap that blade as well. He considered returning the knife a second time, but decided it would be taking advantage of the manufacturer's willingness to pay for his chronic misapplication of their equipment. He still wonders whether anyone at Stockman really believed him when he wrote that he used it to trim elephant toenails.

***

As his first full year as senior keeper wound to an end, Roger settled into the work more comfortably, confident in his ability to get the job done. He bid farewell to Robbins and Cochran, welcomed a succession of new partners, and endeavored to create a workplace democracy rather than a dictatorship. By fostering an atmosphere where dissenting opinions could be expressed without fear of retribution, he acknowledged his team's intelligence and experience. If a troubleshooting session was successful, he made certain they received all the credit. If it wasn't, he accepted full responsibility as senior keeper and shouldered any fallout from management. It became a running joke, with serious implications, that in order to survive in the elephant barn you had to trust your partner with your life, your wife, your children, and your paycheck.

Throughout 1971, Thonglaw had mated with Rosy, Pet, and his daughters Me-Tu and Hanako. Roger had no issue with breeding the older cows, but held serious reservations about the inbreeding because any farmer worth his salt knew the dangers. When he voiced his concerns, he came up against what Al Tucker learned a long time ago: some decisions weren't his to make.

"Portland was the Elephant Capital of the World, and I think the dollar signs blinded certain people," Roger remarked about the breeding program in 2016. "Zoos are run by management, not keepers, and if the moneymen say breed 'em, you breed 'em."

During the first eighty years of the twentieth century, twenty-eight Asian elephants were born in North America, nineteen of them at the Oregon Zoo. Of the nine born elsewhere, none survived to its first birthday, compared to fourteen Oregon calves that survived beyond their first year. Given the Oregon Zoo's astonishing birth and survival success rate, the local populace grew a bit blasé about baby elephants, but there was no arguing they still drew crowds, which put money in the coffers. Every zoo accountant in the world understood the magnetic appeal of a calf, and every zoo wanted one. Over time, several facilities successfully bred their cows, but none matched Portland's level of achievement in terms of calf birth and survival.

Elephants take a long time to mature. Cows are between twelve and fourteen years of age before estrus occurs, and bulls are between ten and fifteen when they produce sperm. Either sex may experience early maturation in captivity, but the reason for this remains unclear. An elephant cow is most fertile between the ages of twenty-five and forty-five, after which she enters menopause.

Unlike human beings, it takes more than an urge and a bottle of wine to make elephants copulate. It's been proven that unless the animals genuinely like one another, mating is literally impossible. The bull can attempt to woo the cow, or even try to force her compliance, but if she's not interested, he's out of luck. At the same time, rejecting a bull can be dangerous for the cow; he may injure or even kill her if he doesn't get his way.

In a captive breeding program, where the elephants' choice of mates is limited, an experienced handler must be well acquainted with the various personalities of his animals in order to determine which pairs are likely to find each other attractive. Ideally, the elephants should be allowed to exercise their prerogative without well-intentioned human interference, but even then there's no guarantee of success.

Though Tucker warned him the title of senior elephant keeper carried variable weight depending on the issue and the level of management he was dealing with, it annoyed Roger that his recommendation against inbreeding was ignored.

"Sometimes I was in charge and sometimes I wasn't, and I never knew which it would be," he said. "The best I could do was try to pick my fights carefully."

Every year, as Portland settled into winter, maintaining a decent temperature inside the barn became an issue. Large electrical Markel heaters had been installed against the ceiling to force warm air down, but they only partially succeeded. Frost still built up along the juncture of floor and wall in some of the back rooms. In the winter of 1971-72, Roger called Bill Scott and Jack Marks, who now served as Director Emeritus, and said he needed them in the barn right away. He assured them no one was dead or dying, but there was something they had to see. When they arrived, he led them to holding pen two and pointed at the line of frost growing where the wall met the floor.

"I wouldn't want to stand on that in my shoes and socks," he said. "I'll bet doing it barefoot is the shits."

Marks and Scott felt the floor, astonished to find it freezing cold. Elephants generally dislike the cold (although adults can tolerate up to thirty minutes outside in winter provided there's no wind, the air temperature isn't exceedingly bitter, and there's adequate traction), yet they seem to love snow. Roger's animals would charge about like human children when allowed outdoors for moments of recreation. Hanako appeared to enjoy it the least, but Belle reveled in gathering up snow with her trunk and hurling it unerringly at the gardeners.

Marks immediately ordered bigger heaters to be installed, plus Casablanca fans to move the warm air downward. Even with those improvements, the rooms remained cold. In one of his last acts before retiring from the zoo permanently in 1972, Marks contracted for the floors in the barn to be jackhammered room by room, radiant heat to be installed, and new concrete to be poured. Non-skid rubber mats were laid on top to provide cushion and comfort to the elephants, but they did more than that. Easily hosed, squeegeed, and dried, they became a godsend to the keepers.

When the work was completed, the elephants were allowed in to inspect their renovated quarters. Within minutes, they were all stretched out and gently snoring. Watching them, Roger nodded with satisfaction and gave Marks and Scott a thumbs-up. If the elephants were happy, then so was he.

***

Ever since Roger began at the zoo, it was his habit to join several other keepers for an after-hours shoot-the-shit in the lunch room of the reptile house, where they drank coffee, bitched about their day, gossiped, and flung a bit of bull before heading home.

One afternoon in 1972, Roger was chatting with Gordon Noyes, the two Floessler brothers, primate keeper Dale Brooks, and bear man Harold Meeker when senior reptile keeper Larry Copy arrived. The men exchanged salty, cordially abusive greetings and Copy moved on into the main building to commence feeding his reptiles. He preferred to do this after the zoo closed for the day because certain members of the general public had expressed displeasure at seeing the boa constrictors, pythons, and large lizards consuming their meals whole. Apparently, that side of "the wild" was a tad more visceral than some visitors bargained for.

Reluctant to leave the easy camaraderie of their fellows but eager to get home to families and dinner (or stop by Goose Hollow Inn to down a few beers), the men in the lunchroom prepared to depart. As they gathered up their things and headed for the exit, they heard Copy shout. "One's got me!"

Lunch boxes, jackets, and thermoses hit the floor in a clatter as the men raced inside. They found Copy bent double over the bottom half of a metal Dutch door, his upper body inside an exhibit, one wrist caught in the jaws of an enormous python. The snake, a female identified by Roger as a "hen python," writhed as she struggled to work her muscular coils up Copy's arm so she could encircle his neck or torso and asphyxiate him.

A python's mouth is lined with slick hooked teeth that curve backward toward the interior of its throat. Captured prey easily slides in, but is impaled as it struggles to escape. Caught fast by the downward slope of those teeth, Copy could not get free. Each tooth was a red-hot needle impaling his flesh, yet somehow he retained enough presence of mind to grab the back of the snake's head with his free hand and squeeze the hinge of her jaw. He couldn't force her mouth open, but the pressure kept her from working his hand farther down her throat. Astonishingly calm given the situation—and in true keeper fashion more concerned with the snake's welfare than his own—Copy directed his rescuers to gently and carefully lift the python out of the exhibit.

This was no easy task. Depending on the species, an adult python can weigh as much as two hundred pounds, have the girth of a telephone pole, and reach a length of more than twenty feet. A calm one is an armload; one that thinks itself under attack becomes a daunting opponent. Swearing and grunting with effort, the keepers each grasped a length of silky, powerful, fighting muscle and hoisted the snake into the open, one combative inch at a time. As they struggled, Copy slowly straightened and stepped backward. The instant both he and the snake cleared the exhibit, someone slammed the door closed on its other inhabitants, grateful the python's companions hadn't decided to make a break for it.

Stretching the furious reptile out flat, they lay down on the bucking coils to hold her immobile. Dale Brooks relieved Copy of the burden of holding the snake's head and pried her jaws open. As Copy gingerly worked his mangled wrist off the python's teeth, he told his friends what happened.

Following standard procedure, he'd checked the location of the reptiles in the exhibit before opening the glassed-in top half of the dutch door. He counted three pythons, as expected, not realizing what looked like a snake was actually a coil from another snake's body. Unbeknownst to him, the hen python lay out of sight along the lower half of the door; not to ambush him, as some might think, but simply because that was where she'd wound up in the course of the day. When he leaned to toss in their meal of dead chickens laced with vitamins and minerals, she struck at her dinner and grabbed his wrist instead. It was a complete accident, Copy stressed, as if worried someone might reprimand the snake. She hadn't meant to hurt him.

While someone rushed him to the hospital, the others returned the furious python to her exhibit. After the door was locked and the handle tried several times to make sure it was secure, the men headed home, poignantly aware of how close they'd come to tragedy. If they'd been five minutes faster leaving, or Copy had been five minutes slower arriving, no one would've heard him cry out.

The aftershock hit Roger on the drive home and his hands shook as he lit a cigarette. His first day at the zoo replayed in his mind: how Lorin Floessler cautioned him to check and recheck before doing anything with the animals, and how quickly he'd let the warning slip his mind. Larry Copy was a conscientious keeper with years of experience, yet he'd still made a mistake that could've turned deadly. Roger shuddered to think about his friend's near miss and took the lesson to heart. No one lived a charmed life.

Copy survived his encounter with the python, but bacteria from her mouth caused a massive infection requiring weeks of strong antibiotics to resolve. He returned to the zoo bearing scars that served as a warning to him and his friends for the rest of their lives.

***

The summer of 1972, Tuy Hoa delivered a calf the keepers named Gabriel. The herd responded enthusiastically to the little bull's arrival, but none was more elated than Belle. Although she'd shown no further interest in mating after Packy was born, she reveled at being an aunt and adopted each infant as it came into the barn.

"If there was something happening that Belle didn't like or understand, she'd grab a calf, *anyone's* calf, and take care of it until things got back to normal," Roger recalled in 2015 when asked about her affinity for all calves despite an obvious lack of interest in producing any of her own except Packy. "She took the job of being an auntie very seriously."

Belle had a lot to look forward to in the coming year because Rosy, Me-Tu, Pet, and Hanako had confirmed pregnancies, all sired by Thonglaw. These births would bring the barn census to fifteen. Roger tried not to wonder where he would put them all, and he contemplated asking for more help to be assigned to the barn.

As autumn 1972 moved in, bringing rain and littering the ground with bright patches of leafy color, word came down from Bill Scott that two-year-olds Tina and Judy had been sold. The news struck Roger dumb. He couldn't understand the logic behind getting rid of two perfectly healthy calves that would grow into gorgeous cows. The answer, he later learned, was twofold. First, someone on high had at last expressed concern about inbreeding. Second, it would take at least a decade before Tina and Judy were ready to breed. Better to make room for adult cows in their prime that would add new blood to the gene pool.

Roger dutifully weaned the calves from their mothers, taught them the shalts and shalt-nots of zoo life ("Thou shalt listen to the keeper, thou shalt not step on the keeper's foot"), separated them from the herd, and placed them in quarantine—standard practice for any animal entering or leaving the zoo to prevent the spread of disease. On September 15, 1972, Judy boarded a truck headed for Wildlife Safari in Winston, Oregon. A few days later, Roger learned the crate they intended to use to ship Tina to her new home at Vancouver Game Farm in Aldergrove, British Columbia, was too big to fit through the doors of the cargo planes that landed at Portland International Airport. Instead, he and keeper Roland Smith would drive her over the border to an airport in Canada.

They'd already worked a full day when they climbed into a beat-up rented van and set out on the five-hour journey with Tina in her crate riding in the back. Smith was a hard-working kid much younger than Roger, with black hair, a beard, and the rugged good looks of actor Burt Reynolds. Roger liked him for his intelligence, initiative, ability with animals, and sense of humor. He figured with "Rollo" along for the drive, at least the trip wouldn't be boring.

That was the case for thirty miles or so, until the truck's muffler fell off somewhere around Woodland, Washington. Unable to stop for repairs—after all, what would they do with the crated elephant—they drove on. The engine's roar made it impossible to talk or listen to the radio, so they just stared through the windshield as the miles spooled along a seemingly endless highway. Roger hoped poor Tina, stuck in the back in the dark, didn't think the world was coming to an end.

Nearly three hundred miles later, they approached the checkpoint at the border. A stern-faced guard looked at their papers and gave the tired men a once-over with a jaundiced eye. "Says here you've got an elephant. How the hell did you get an elephant into a truck this small?"

"Well, sir, she's not a big elephant," Roger said with extreme politeness. Thinking he was being made fun of, the guard yanked his flashlight from his belt and brusquely ordered them to raise the truck door so he could inspect their cargo. As the door rolled up, Tina released a loud squall. The border guard leapt backwards and nearly dropped his flashlight. Running the beam across the crate, he picked out the gleam of one eye and the coil of her inquisitive trunk. "I'll be damned! That *is* an elephant!"

"Did you honestly think I'd lie about it?" said Roger.

They invited the guard inside the truck to meet Tina and feed her an apple, then continued on their way. Thirty miles later, they pulled up beside the airport freight terminal and cut the engine. Reveling in the blessed silence, they swiftly unloaded Tina's crate, crooning to the anxious calf as they worked.

Nearby, six horses were also crated and awaiting transport. They stared at Tina, mystified, presumably never having seen or smelled anything like her. The first time she trumpeted, sounding a bit like a tin horn, the horses shied, whinnying in terror, and dropped a load of manure.

The horses' owner shot dirty looks at the elephant and her keepers as he tried to calm his animals. Finally, Roger turned to him, lifted both hands in the air, and shrugged. "Well, shit, I can't muzzle her!" The equestrian stormed off in a huff.

Roger and Smith raked manure from Tina's crate, offered her as much food and water as she wanted, and lavished her with love and attention. As the moon rose, they bedded her down for what they hoped would be an uneventful night. Overcome with foot-dragging exhaustion, they sought their hotel room and the oblivion of sleep.

Rousting out of bed early the next morning, they made themselves presentable with showers and clean uniforms before returning to the airport. Tina greeted them with joyful trumpeting, inciting the horses into another nervous breakdown. Not bothering to hide their amusement, Roger and Smith tended to Tina's needs one last time and kept her company until her flight. When she was loaded and secured, they stayed to watch the airplane depart. As it taxied down the runway and lifted into the sky, Roger swabbed his streaming eyes with a handkerchief. "Damn dust," he muttered, and loudly blew his nose.

He took the story home to family as he regularly did, entertaining them with tales from the zoo. His notion of family broadened over the years to include several "adoptees," friends of his daughters, along with some of the newer keepers who shared his ardent devotion to animal welfare. While Roger participated in regular gatherings of The Gourmet Poker Club, a group of six elephant program regulars who took turns hosting, RoseMerrie invited the female keepers to dinner at least once a month in a show of feminine solidarity.

On January 9, 1973, Me-Tu delivered a male calf the keepers dubbed Stretch for his long legs. Though Rosy was his grandmother and should have had first dibs on the infant, Belle made it a point to smugly appropriate him whenever she got the chance.

Five months later, on June 17, Pet gave birth to Stoney, so named because he looked like a 225-pound rock. On July 19, Circus Vargas took ownership of one-year-old Gabriel. Roger disliked having to separate the young elephant from his mother, but didn't push too hard on the issue. He understood the logic behind selling Gabriel even if he didn't like the timing. The spot left vacant by the calf's departure could be filled by a cow, almost always the milder-tempered of the sexes. The little fellow was at least a decade away from musth, but it would happen.

Adult bulls were notoriously difficult to keep and many zoos refused to even try. Many facilities chose to put down their bulls as the elephants approached musth. Portland having two in residence was definitely an anomaly and something of a wonder. Maintaining a bachelor herd was thought to be impossible and foolhardy, but that viewpoint is currently undergoing a sea change. The Denver Zoo, Houston Zoo, and Birmingham Zoo now support bachelor herds. Denver has five Asian bulls, Houston has four, and Birmingham has the distinction of being the first accredited facility in the United States to successfully recreate a herd of African bull elephants. It was once believed that male elephants in the wild led a solitary existence except when breeding, but that has been proven incorrect. Bachelor herds do occur naturally, and that knowledge has encouraged the development of the first captive bachelor groupings.

Roger had experienced five elephant pregnancies and births since coming to the zoo, but it never became commonplace. He fretted through gestation and hovered like an anxious father-to-be during delivery. As each pregnancy neared term, he made it a point to put the expectant mother with her closest friends in the herd, those who would best offer emotional support during birth and babysitting services afterward.

In August of 1973, thirteen-year-old Hanako began the days-long process of delivering her first calf. Because she was a nervous and unpredictable animal, Roger chose her mother, Tuy Hoa, and Belle as midwives. Tuy Hoa's presence would help calm Hanako, and Belle's position as matriarch would lend an air of trust and control. Inexperienced first-time mothers had been known to attack and even kill their offspring. These were usually cows who lived in isolation, had never benefited from the presence of older, more seasoned females, or had grown up without observing the birth process first-hand. Hanako had repeatedly witnessed such events, but still Roger worried because, as he put it years later, "She could be a total dingbat."

When the birth process begins, the cow passes a large rubbery blob of mucous called the cervical plug. Colostrum, a sticky yellowish liquid rich in immune factors, may leak from her breasts, and there is some bleeding from the urogenital area. Birth occurs from one to five days after the plug's release, and happens quickly once the mother enters active labor. After the calf is on the ground, the mother uses her trunk and front feet in an unexpectedly robust manner to rouse the infant as rapidly as possible, usually within five to fifteen minutes. This unnerving spectacle can look like an attempt on the calf's life, but is exactly the opposite. This is instinct at work: just as a wild mother needs a watchful and protective midwife to prevent predators from snatching her defenseless newborn, she also needs the calf to stand up and be able to run from danger as soon as possible.

Hanako's labor progressed smoothly and she delivered without incident. The calf was on its feet soon after it hit the floor. That's when trouble began. The hungry infant understood who momma was and what it needed from her, but Hanako refused to let it nurse. Instead, she behaved as if she had no idea where the little stranger came from or what it wanted. To escape the calf's persistent advances, she paced the maternity ward in endless circles, striding to keep ahead of the ungainly infant as it toddled after her in vain pursuit. Meanwhile, Tuy Hoa and Belle watched the show but did nothing to intervene.

Roger shook his head as he watched. Hanako had always been a bit odd, what he called "a half-bubble off." He should have expected issues with her parenting skills. Somehow, he thought Belle and Tuy Hoa would handle any problem, but they stood there like they had nothing better to do. "Okay, lads," he said to his crew. "Let's get that calf out of there and get it something to eat. We'll bring it back to her later, see if she's more interested then."

The zoo's newest veterinarian, Dr. Michael Schmidt, cleared his throat and suggested they leave rather than remove the calf. He believed Tuy Hoa and Belle weren't getting involved because they viewed the keepers as the ones in charge. They were waiting for the men to either do something or get out of their way.

His reasoning made sense, but did nothing to settle Roger's nerves. Still, they'd lose only a few minutes by giving it a try and they could always take the calf to the nursery afterward. Working quickly, they closed the barn and darkened the exhibit room until the only illumination came from the night lights. Retreating into the breezeway, they hunkered down on the cement floor where they'd be out of sight, but could still peer around the corner to monitor what occurred. There they waited, silent and expectant.

Belle and Tuy Hoa likely weren't fooled by their departure. An elephant can hear another of its kind over two miles away and their sense of smell is excellent. They knew the keepers were still there, but they understood the field was cleared to let them go to work. In what later became known as "The Great Elephant Ping Pong Game," the two cows converged on Hanako from opposite sides and bounced her off the back wall with their heads, trunks, and shoulders. The sound of her hitting the concrete set Roger's teeth on edge. Hanako trumpeted in protest and tried to escape, but the older cows kept at her until she cowered in submission and allowed her exhausted newborn to latch onto a nipple.

Shoulder to shoulder, Belle and Tuy Hoa bracketed Hanako, refusing to let her move. She stood with her head bent in defeat. After a moment, she hesitantly reached out with her trunk and touched her child, exploring its body and drawing in its scent. A rumble of contentment sounded through the barn.

Roger sighed and sagged back against the wall. He looked over at Schmidt and tipped his hat brim by way of acknowledgment. The new vet's instincts had been spot-on. Unfortunately, their success was ultimately a hollow one. The calf died after three days, succumbing to what was later diagnosed as white muscle disease, also known as nutritional muscular dystrophy, a disorder caused by a deficiency of selenium and vitamin E.

Unlike poor Hanako, Rosy was an old hand at this pregnancy business and sailed through her latest like a wide-bodied ship on calm seas. On Halloween 1973 rather than help RoseMerrie shepherd their costumed daughters around the neighborhood, Roger was in the barn to witness the arrival of the zoo's fifteenth elephant calf. Though he sometimes kept odd hours, Roger did his best to be home most evenings for dinner. He made it his duty to keep the cookie jars full of snickerdoodles, and he almost never missed reading to seven-year-old Michelle and four-year-old Melissa before bed. On those occasions when he couldn't be there because of something happening with the elephants, nobody minded. All three of the women in his life knew he was doing important work.

When the newborn arrived, the midwives seemed more interested in Rosy. The calf lay where she'd dropped it, unmoving inside the amniotic sac. Usually, the birth sac ruptures when the infant hits the floor, but this one hadn't, and the cows appeared disinclined to tear it open. Roger anxiously waited for what seemed like forever, but was probably only ten seconds, then rushed in and ripped the sac with his hands. The calf immediately rolled over and looked at him. With a flush of paternal delight, he named the little female M&M, in honor of his daughters.

Chapter *Seven*

A KEYSTONE SPECIES

1974–1977

A man of simple desires, Roger never wanted much beyond the ability to provide for his two families—human and elephant—enjoy the company of friends, do some gardening, and take a yearly hunting trip. The thought of awards or accolades never occurred to him, so he was entirely flummoxed when word reached him that he'd been chosen to receive the 1974 R. Marlin Perkins Certificate of Excellence from the American Association of Zoo Keepers (AAZK). (Not to be confused with The R. Marlin Perkins Award for Professional Excellence given by the Association of Zoos and Aquariums.)

Perkins was a zoologist and advocate for the protection of endangered species. He'd served as director of the Buffalo Zoological Park (now the Buffalo Zoo), Lincoln Park Zoo, and the St. Louis Zoo, and in 1960 he joined Sir Edmund Hillary's expedition to the Himalayas in search of the yeti. His greatest claim to fame, however, and what he's most widely remembered for, is being the host of *Mutual of Omaha's Wild Kingdom*. Beginning in 1963, the television program brought a glimpse of nature into American living rooms, and provided a first introduction to the concepts of conservation and ecological responsibility.

Roger and RoseMerrie flew to Chicago to attend the AAZK annual conference in April 1974. The award, given in recognition of "professional attitude, true dedication, superb application of animal husbandry practices, and your contributions to the welfare of the animal life placed in your charge," left Roger humbled and speechless. As far as he was concerned, he'd done nothing except the job for which he'd been hired.

"In truth, I felt like a fraud," he said in 2016. "Everything I knew, I learned from Al Tucker. He should have been the one up there shaking hands, not me."

Back home, he got to thinking. Not about the award, per se, but about the feelings and motivations behind it. Slowly, his own nebulous ideas about conservation began to develop and evolve, beyond the basic notion that every species is worth preserving. He searched out newspaper and magazine articles and read books. He began to dream that, in the future, all zoos in the world might come together and each choose one or two endangered species on which to concentrate their protection. All these years later, he remains dedicated to the preservation of the natural world. "We have to do something," he said in 2016. "To try and fail is forgivable, but to be so indifferent that you never try is immoral."

Elephants, of course, were Roger's top priority, and the more he read about them, the more he realized what a vital position they hold in the natural world. Elephants are what's known as a keystone species, an animal capable of modifying its habitat to the benefit of others. In Kenya, for example, elephants in search of salt have hewn vast caves in the side of Mount Elgon; their immense excavations provide shelter and salt access to other animals in the region. Over two dozen types of trees rely entirely on the elephant for seed dispersal. Not only do elephants plant trees, but they also fertilize and prune them. By opening forest land, elephants create natural firebreaks and grassland for grazing herds and the animals that hunt them. Their digging provides water access, and their manure dispenses essential nutrients into the soil; they are the gardeners of this planet. The most important thing to remember about a keystone is this: remove it, and the entire arch begins to crumble.

Two hundred years ago, the population of wild Asian elephants was estimated at 200,000. By the turn of the twentieth century, reports from across the thirteen range states (Bangladesh, Bhutan, India, Nepal, Sri Lanka, Cambodia, China, Indonesia, Lao People's Democratic Republic, Malaysia, Myanmar, Thailand, and Vietnam) placed the number at 39,500–43,500 animals, plus an additional 15–20,000 in captivity worldwide. Other estimates quote a much lower number, suggesting there are less than 35,000 total. By 2015, the population of Sumatran elephants, the smallest of the Asian species, dwindled to less than 3,000. A degree of uncertainty shadows these numbers because of difficulty obtaining a census due to dense vegetation, difficult terrain, and sometimes outmoded survey techniques, but the indisputable fact remains: elephant populations are steadily declining worldwide. The three greatest threats to elephants are habitat loss and fragmentation, human/elephant conflict, and predation. The root cause of all three is humankind.

Intensive logging, the clearing of land for agriculture and livestock, and the rampant spread of human settlements have steadily eaten into land once belonging to the elephant, breaking and blocking ancient migratory routes. The resultant separation of herd from herd, and wandering bulls from receptive females, hinders the passage of knowledge and expertise to younger animals and creates a crippling loss of genetic diversity. It adversely affects the socialization of young elephants, leaving them bereft and clueless. Without guidance through the long years of maturation, they may end up looking like an elephant, but they'll never achieve the greater measure of what it means to *be* an elephant. In the early 1980s, a group of teenage male African elephants, survivors of a herd culling (killing) at Kruger National Park in South Africa, were relocated to Pilanesberg National Park, more than five hours away. In 1993, rangers at Pilanesberg discovered several of the park's rhinos mutilated and killed. The rangers subsequently learned that without the presence of older, more experienced bulls to influence and teach them, the young male elephants—who were suffering the effects of PTSD brought on by the cull a decade earlier in which they witnessed their families being killed—were on rampage. These elephants entered full musth ten years earlier than normal, and also simultaneously, something never before documented. When adult males were eventually introduced to the area, the teenagers flocked to them and the killings stopped. This illustrates an important role older bulls play in elephant society beyond the breeding imperative: younger bulls need them as role models.

Elephants cut off from their annual passage to areas of plentiful food quickly over-browse an area. In search of sustenance, they raid cultivated fields. The loss of revenue to large agricultural ventures is estimated to be in the millions of dollars. Small-scale farmers have watched elephants wipe out their entire livelihood in a single night. Attempts to protect arable land with ditches or high-voltage fencing have failed because elephants will fill ditches with dirt and branches in order to walk across them, or will push trees onto electric wires and clamber over the fallen trunks. Since ivory doesn't conduct electricity, bulls (and cows, in the case of African elephants) use their tusks to tear out fences, holding their trunks out of the way. Their spongy feet are also poor conductors of electricity, which allows the elephants to simply trample fences.

Desperate for survival, the farming families—many of which belong to cultures that revere the elephant—may attempt to drive the marauders away by lighting fires, making noise with drums and other implements, setting off firecrackers, or shooting guns into the air. Even when these methods work, they're usually nothing more than a temporary fix. The elephants soon return, at which point the farmers—unwilling to choose an elephant's life over that of their child or to sink into an even deeper well of poverty—see no alternative but to kill the problem animals, sometimes as many as one hundred annually.

Conflicts between people and elephants can turn deadly for humans, too. In India alone, elephants kill roughly one hundred people each year; some years, the count may be as high as three hundred. Half of those deaths are the result of chance encounters.

Innocent animals living apart from human settlements, or as far afield as they can get, suffer as well, hunted for their meat, hides, and ivory. Herds are sometimes culled to free up land for agriculture or to capture calves for black-market wildlife sales. Lured by the call of easy money, some poachers prey on the very animals that are interwoven into their cultural traditions. In 2012, poachers armed with grenades and AK-47s slaughtered more than three hundred elephants in a single day at Bouba Ndjida National Park in Cameroon.

The more Roger learned, and the better he came to realize that elephants could actually disappear from the earth in his lifetime, the more determined he became to do something. He wasn't sure how he could help wild elephants survive, but in Portland he could fight to give zoo elephants the best and longest lives possible.

During the spring and summer of 1974, it seemed as if a revolving door had been installed in the barn. Over the course of three months, one-year-olds Stoney, Stretch, and M&M were sold to outfits in Florida, Canada, and Texas, respectively. In June, Roger drove to California to pick up a hand-raised four-year-old Asian elephant female named Tamba from the Oakland Children's Zoo, where she'd lived alone without another elephant. Isolation had made her shy, gentle, sweet-natured, and completely focused on human beings. Roger took an instant shine to her and apparently the attraction was mutual. She began following him around whenever she could, mooning at him with big eyes and begging for attention.

"She was in love with me to the point of my embarrassment," Roger recalled in 2015. Several attempts were made to integrate her into either Rosy's herd or Belle's, but Tamba never fit in well with the other animals or developed any close relationships. The Oregon Zoo elephants had been divided into separate herds to accommodate the need of both matriarchs to lead and forestall any major arguments between them.

However, Tamba was not without human friends. Gordon Noyes, a prior elephant keeper who'd moved on to become senior keeper of bears, became a frequent visitor during his divorce. He'd sit in the barn, playing his harmonica for Tamba with the door to her enclosure cracked so the two of them could reach through to each other. It was simultaneously the funniest and the most heart-breaking thing Roger had ever seen.

Zoo administration decided Tamba's disposition made her the perfect candidate to provide rides to children, at least until she grew old enough to breed. They built a howdah—a platform with a railing strapped onto an elephant's back—and with help from Director Ogilvie's eight-year-old son, Billy, Roger trained Tamba to walk quietly with passengers. She performed for two years without incident, until administration replaced Roger with an inexperienced handler. Less than a month later, Tamba broke from the woman and ran, fortunately back to the loading platform rather than across the zoo. No one was hurt, but when she escaped a second time the rides were permanently suspended.

Twenty-three-year-old Susi arrived in early November 1974 from the Ralph Mitchell Zoo in Independence, Kansas, where she'd lived alone for eighteen years. The reason they chose to get rid of her remains uncertain. They were anticipating the arrival of a four-year-old calf, so elephants in general were no problem. Perhaps they didn't have the resources to care for two animals and knew a playful calf would draw larger crowds, but an advertisement offering Susi for the price of $500 due to rock throwing offers an illuminating clue. The Oregon Zoo chose to take her because she was of breeding age and, as far as anyone knew, unrelated to the rest of the herd.

The first time Roger saw her, his paternal instincts shot to the surface. "There wasn't a bad bone in the poor dear," he said, speaking in 2015, "but she was so damned ugly she hurt your feelings."

Like Tamba, Susi never adapted to life among her own kind. After all those years alone, she had no idea how to properly respond to their friendly overtures and spent her days standing apart, content to be in the presence of other elephants without drawing attention to herself. Being among them obviously soothed something in her soul, because she never threw rocks again. At mealtimes, she'd wait until the other elephants were busy feeding, then she'd pick up a foot-thick slab of hay with her trunk, balance it on top of her head, and carry it to the far end of the yard to eat.

Roger described it as "the damndest thing to watch," and never figured out why she did it. The other elephants weren't giving her trouble or stealing her food. "I guess she was used to eating alone and preferred it that way," he said.

***

A few weeks after Susi's arrival, Roger rolled a heavy wheelbarrow across the yard and upended the load onto the manure dump. The morning's high of fifty-three degrees was slowly dropping, and he could feel the threat of frost in the air—not surprising with Thanksgiving three days away. As usual, RoseMerrie had the holiday well in hand, and Roger looked forward to the raucous gathering of family. The kids—his own girls, plus nephews and nieces—would watch the Macy's Thanksgiving Day Parade, and the appearance of Santa Claus at the end would ring in the screaming mania of Christmas before they'd even taken the turkey out of the oven. He'd have a few drinks with the other guys, they'd all eat until they couldn't move, and then they'd fall asleep in front of the television watching football. The day would be the perfect antidote to the worry that persistently dogged him: the vague sense that he was letting down his family by spending so much time at the barn, that he wasn't doing enough for the elephants, and that there must be more he could accomplish if only he were smarter.

Much of what he'd initially found intriguing about elephants—their sounds, their smell, their quirky behavior—still held his interest, but had drifted into the background and become barely noticeable, supplanted by the sheer volume of daily work required to properly care for them. There were eleven elephants at the moment, and Roger felt a little like Sisyphus; every day, he rolled a ball of bolus to the top of the hill, only to have it chase him back down. Occasionally, he longed for the old days when he'd been a floater—busy, but with an opportunity to take an occasional break to visit with the elephants, maybe bestow a bit of affection. These days, there was barely time to take a piss.

He glimpsed Morgan Berry heading into the barn to see Thonglaw. He didn't approve of Berry's methods and never would, but the fact remained that Thonglaw's former owner was the only one who could tend to the bull's medical issues, particularly his feet.

Roger would never forget his own first close encounter with the bull, but neither had he held it against him. Truth was, he admired the cantankerous elephant and even felt a certain wry, tolerant affection for him. One couldn't hate an animal for being true to its nature—least of all one whose personality had been molded by human contact.

He caught up with Berry inside and found him deep in conversation with Dr. Schmidt. Rather than interrupt, he joined the keepers gathered near Thonglaw's enclosure. The twenty-seven-year-old bull stood against the far wall, broad forehead pressed into a corner, either unaware of their presence or, more likely, ignoring them. Body hunched in an attitude of pain, he made no sound or movement.

In his younger days, Thonglaw had been magnificent: at least ten feet tall at the shoulder, with a beautifully shaped head and body, a gorgeous pair of tusks, and attractive mottling along the bottom of each ear which his son Packy had inherited. He'd stride around his enclosure, king of the world. By his early twenties—not quite middle-age for an animal whose average life expectancy is forty-five—he'd developed a stiff front leg. This may have been due to arthritis, a congenital abnormality, an undetected injury, conditions of a captive life, or elements of all of these. The frozen joint affected his gait, which caused irregular weight placement on his other feet, leading to abnormal growth of toenails and soles.

Because his hostility made it impossible for keepers to provide him with regular care as they did the cows, Berry was called in on several occasions to force resentful compliance when hands-on treatment was required. By Berry's own admission, he'd gotten busy the past few years and hadn't tended to the bull's needs as diligently as he should have, and now Thonglaw's feet were infected. His pain and lameness were so great he could no longer mount the cows to breed. He was losing weight and growing weak, and he often refused to walk at all. Tranquilizer darts were used in the past to subdue him, but the drugs were unpredictable in those early years, particularly with large animals, so were used sparingly and with extreme caution. Putting an elephant down for a protracted length of time can kill it. Unlike other mammals, an elephant has no pleural cavity—the narrow, fluid-filled space between the membranes of the lung and the inner chest wall. Instead, an elephant's lungs are anchored to its ribcage. Rather than inflating and deflating the way human lungs do, elephants depend on their chest muscles to do the work. If they lie down for too long, their chest becomes constricted, and they can have trouble breathing or suffocate.

Thonglaw glanced around as Berry entered the enclosure and spoke his name, but there was no fight in his flat gaze—an indication of how poorly he felt. Dr. Schmidt injected the bull with a light dose of acetyl promazine, a common tranquilizer used to quiet aggressive animals. Slowly Thonglaw sank to his knees and stretched out on one side. The second he was down, Berry, Roger, and the barn crew went to work excising affected foot tissue and applying medication to fight infection and pain.

The operation went beautifully. Thonglaw tolerated both the drug and the procedure, and Roger felt a stir of hope. If this worked, caring for Thonglaw wouldn't be such a chore. There was a real possibility the issues with his feet could be resolved and the bull would become his old virile self again.

With the work finished, everyone retreated to a safe distance while Schmidt injected a drug to reverse the tranquilizer. Not long after, Thonglaw stirred, shifted onto his chest, and hefted to his feet without incident. As he headed toward a pile of fresh hay, a soft cheer rose from those assembled.

Suddenly, the bull collapsed, falling with a crash that shook the floor beneath Roger's boots. Berry, the first to reach him, cried out that Thonglaw had stopped breathing.

The crew scaled the immense body and jumped up and down along Thonglaw's ribcage in an attempt at CPR. "If we'd had a compressor with a hundred-gallon tank we might have blown enough air up his nose to do some good," Roger said, thinking back to that time. "But five-hundred-plus pounds of us trying to spring his ribs to flex his lungs was useless. I felt so sorry for Morgan."

Berry always insisted he loved Thonglaw despite the training methods he'd employed. He'd purchased the six-year-old in Bangkok in 1953 and brought him to the United States. He'd nurtured him, trained him, performed with him, bred him, and berated him. Now the elephant lay dead beneath his hands. Berry would later tell a reporter that when Thonglaw perished, something inside him died as well.

A post-mortem failed to reveal an underlying cause of death, and staff was left to assume the legendary sire had suffered an abnormal reaction to one of the drugs used. The herd was somewhat subdued in the days following his death, likely sensing through infrasound that he was gone, but his absence seemed to be lightly felt among his harem.

Not so for the keepers. Thonglaw had been a constant test of their fortitude, persistence, patience, and innovative ability, but he'd also been a source of pride, admiration, and awe. When they learned it was their responsibility to "dismantle" him, as Roger delicately put it, they were horrified.

"The sad truth is that everything that lives dies, and you have to do something with what's left behind," Roger said more than once during interviews. That reality had been driven home to him during his boyhood on the family farm, and had only been reinforced during his time working at the zoo, surrounded by animals. "I told my crew they'd better damn-well be there and not put up a fuss about it or I'd escort them out of the barn on the toe of my boot and they'd never come through the door again. Like it or not, it's part of the job, and you don't send old friends out alone." These days, any elephant that dies at the Oregon Zoo is trucked to a heavily wooded zoo-owned property east of Portland and buried. Zoo staff follow Department of Environmental Quality guidelines to prepare the gravesite and disinfect their equipment after the burial.

As a hunter, Roger had field dressed the game he killed, but this was unimaginably worse. His brain struggled to accept that he was cutting up someone he knew. Thonglaw was not a friend of Roger's—it was difficult to be friends with an animal who wanted to kill you—but Roger certainly admired him. Pulling on coveralls and gloves, he set to work, gritting his teeth against the roil of his stomach, closing his nose to the smell of blood and viscera, and his ears to the whining cry of chainsaws working their way through bone.

Thonglaw was buried in an isolated spot on zoo grounds. Before being interred, his tusks were removed from his skull so they could be displayed as a memorial to the famous bull. Carrying the smaller of the two into the front area of the barn that day, Roger was keenly aware he was handling more than forty pounds of prime ivory that an unscrupulous person might find tempting. As a precaution, he slid the tusk out of sight between several bales of hay. According to another keeper who was there that day, the larger tusk was left on the east side of the hall.

When the grisly job was done and the area cleaned and disinfected, Roger sent everyone home. They'd clocked more than twenty-four straight hours since arriving the morning before, and they couldn't manage an additional minute. He locked up the barn, bid the security guard goodbye, and drove home beneath gray skies.

The house was warm and blessedly quiet with RoseMerrie at work and the girls at school. Roger was grateful for the lack of questions. So physically and mentally exhausted he couldn't yet cry, he downed a mouthful of cheap rye, grabbed a pillow from the couch, and stretched out on the floor in his stinking barn clothes. He fell asleep in an instant and probably would have remained so for hours except the telephone rang.

It was the security guard from the zoo. The larger of Thonglaw's tusks was missing from the barn. Half dazed by lack of sleep, Roger hurried back to the zoo.

According to a newspaper account, someone with a key to the elephant barn removed the tusk between 3:30 p.m., when a workman last saw it, and 4:10 p.m., when the security guard reported it missing. Police were summoned and an investigation ensued. As the senior keeper and the last to leave the barn that day, Roger became their prime suspect. At the time, the accusation made him bristle. Now, he concedes their suspicions were valid, if misplaced. All zoo personnel were questioned, but no leads on the theft emerged.

The specter of Thonglaw haunted Roger and his crew, making them work even more diligently to keep foot rot at bay, but pododermatitis is insidious. Once it gains hold it's difficult, if not impossible, to eradicate.

Roger was particularly concerned about Packy. The last thing he wanted was a replay of what had happened with Thonglaw. No longer a tractable little calf, the teenager had risen to the rank of breeding bull as his sire's vitality waned. In the wild, where competition might well have been fierce, Packy would not have reached a state of sexual dominance until his twenties. At the zoo, with no challengers to his supremacy, he was primed and ready, and had already been bred to half-sisters Me-Tu and Hanako. He stood over ten feet tall at the shoulder, weighed almost seven tons, and his episodes of musth were regular and impressive. Only a fool would enter his enclosure, and that left his foot care in question.

In early 1975, Roger went to Director Ogilvie and petitioned for funds to build an elephant restraint device (ERD), loosely based on the V-shaped timber construction called a crush, which had been used for centuries to subdue elephants in logging camps in Burma. "Crush" is a misleading term; the device in no way crushes an elephant, but safely holds it immobile, allowing keepers and veterinarians free access to treat ailments without putting themselves or the elephant at risk. Roger envisioned a hydraulically powered system with thick walls of widely spaced bars capable of sliding and pivoting to match each elephant's unique shape and size.

Ogilvie agreed to present the idea to the governing board, but the ERD was deemed too expensive to build. Not long after, Ogilvie resigned. (Twice in his tenure, he'd brought the zoo to bankruptcy.) When Roger presented the ERD proposal to the zoo's new head honcho, Warren Iliff—formerly of the National Zoo in Washington, D.C.—he found a willing and eager accomplice in elephant care. Iliff promised to do what he could to make the ERD a reality.

A kindhearted soul with a soft spot for elephants and chimpanzees, Iliff began the process of remodeling the outmoded Oregon Zoo almost from the moment of his arrival in 1975, naturalizing the exhibits into better homes for its animals. He and his wife Ghislaine "Gigi" Iliff were so deeply concerned with the animals' welfare, they often arrived at the barn in the middle of the night to help care for any young elephants in need of extra love. On Christmas, they came in to help feed the animals so those on duty could finish early and go home to their families. Warren's generosity of spirit was so great that when Charlie the chimpanzee bit off the end of his left middle finger, he promptly forgave the ape and reportedly said, "He's a nice guy. We had lunch."

Quickly dubbed the "idea-a-minute man," the new director often lurked near the time clock in the morning to waylay whoever was the target of his latest scheme for improvement. The moment his quarry came into view, he'd fall into step beside them and talk them clear across the zoo if need be, barely pausing for breath.

On one occasion, Iliff collared Roger and brandished a five-dollar bill. He explained that an elderly woman who volunteered as a docent had given him the money to buy treats for the elephants. Smiling, he slipped the bill into Roger's shirt pocket and strolled away.

Roger stared after him. In those days, five dollars could buy quite a lot, but not enough of anything to split among a herd of elephants. What the hell was he supposed to do with the money?

On his way home that afternoon, he stopped to purchase a pack of cigarettes. While standing in line to pay, he noticed a counter display of Horse Shoe chewing tobacco on sale for fifty cents a package. Inspiration struck and he purchased ten packages. The following morning, he quartered each plug and halved the quarters. Lining the cows up along the rail, he ordered, "Trunk up!" and went down the row sticking tobacco into each eager mouth. The elephants chewed with gusto, eyes half-closed in bliss, relishing the sweetness. Tobacco became a regular treat, but given in such small amounts that the possibility of health concerns was insignificant. (These days, the Oregon Zoo herd receives tobacco only on very special occasions.)

Roger never met an elephant that wouldn't turn itself inside out for tobacco, but Pet was the most avid. One day, she followed close behind him as he worked. He thought at first she just wanted company, but the moment he stopped to rake a tobacco wad out of his cheek with an index finger, her trunk flew to the center of her forehead and she commenced to squeak. Laughing, he gave her the worn-out plug and treated himself to a new one. The two friends stood together, chewing amiably until he went back to work.

Hanako also enjoyed the treat and Roger was happy to provide it, hoping to sweeten her disposition and also to dispense a little extra tender loving care to her during her pregnancy. She'd lost her first calf to disease, and while she hadn't appeared to mourn, he knew elephants remembered such things. As the months passed and the fetus inside her grew, he fervently hoped the calf would arrive safe and healthy. In the early days of 1976, as time narrowed toward her projected due date, Roger again placed her with Belle in hopes the older cow's confident demeanor would provide a calming influence.

On a mildly rainy and windy day in February 1976, before an audience of keepers, veterinarians, and staff, Hanako's second calf arrived. Delivery was textbook, just as Roger had wished for, but the moment the newborn female struggled to her feet for the first time, everyone knew something was terribly wrong. Calves are wobbly at the beginning, but nature has hardwired them to gain control quickly. Not only was this calf uncoordinated, but she appeared listless and disinterested in her surroundings. The staff wanted to get their hands on her to find out what was wrong, but hesitated to enter the enclosure. Hanako's prior disinterest in her offspring had shifted 180 degrees, transforming her into a fierce and protective mother. She watched the men warily, and Roger wondered if she somehow blamed them for the death of her first calf.

Eventually, the adult elephants dozed off with their foreheads against the exhibit room bars, the calf slumbering on its feet between Belle's front legs. As bystanders watched from the keeper alley, the newborn's head slowly drooped until it slid neatly between two of the bars and hung outside the cage. Built long before elephant breeding was a gleam in Portland's eye, the barn had been constructed to house only adult animals. The bars were spaced so a person could slip between them, but no thought was given to what might occur with a newborn elephant. Chains were eventually installed between the bars to corral any wayward youngsters, but before then, several calves escaped to lead keepers on a merry chase through the back areas of the barn.

The calf's head sagged lower; she was clearly sound asleep. Reckoning the infant probably weighed only slightly more than he did, Roger told those assembled what he intended to do, then grabbed the calf behind the ears and yanked her into the alley.

She woke screaming in terror, her cries bright and piercing. Belle and Hanako flung their heads up in alarm, roaring deafeningly as they tried to sort out what was happening. Realizing his window of opportunity had abruptly closed and maybe this wasn't such a grand idea after all, Roger proceeded with his plan rather than give up. As he hauled on the panicking calf, it came down on his left foot. Roger went over backward, landing hard but still hanging on. Overbalanced by his grasp on her ears, the baby elephant fell across his legs, pinning him to the floor.

Hanako bellowed in outrage and struck Belle with her shoulder, knocking the matriarch aside. Ears fanned, she leaned against the bars, thrusting her trunk between them in search of the man who'd dared to lay hands on her child.

Icy terror washed over Roger. Time lagged, deadening sound and turning swift action into slow motion. On the periphery of awareness he could hear people shouting, but their words made no sense. Someone clutched his jacket collar in a death grip, choking him as they tried to drag him backward despite the dense weight of the struggling, screaming calf across his thighs. The simple thing would have been to release the calf and let her return to her rampaging mother, but she clearly needed attention and they'd already waited so long to get their hands on her.

Roger tightened his grip.

Over the mound of the calf's hairy body, he watched, helpless, as the tip of Hanako's questing trunk inch closer to his pant leg. If she managed to catch hold, she'd drag him inside the cage, shattering his hips and shoulders against the bars, and press him to the ground with her great forehead until he was dead—which mercifully wouldn't be long. Staring into the elephant's wild, angry eyes, he thought longingly of RoseMerrie, Michelle, and Melissa and waited for the end.

Trumpeting in alarm, trunk swinging in wild arcs, Belle rammed Hanako with her head. The younger cow staggered sideways, bawling in protest as the matriarch stepped between her and Roger. When she attempted to shove past, Belle hit her again, hard enough to turn her broadside. Shoving and swatting, using shoulders and head, body and trunk, Belle drove the frantic mother across the cage to the rear wall and held her there until the barn crew pulled Roger to safety and took custody of the calf. As soon as her child was out of sight and earshot, bound for the nursery, Hanako quieted. The two elephants stood together, breathing heavily. Belle murmured soft noises of reassurance and stroked Hanako's face with her trunk.

The entire event lasted less than a minute, but felt much longer. Roger lay where he'd been dropped, heart thundering in his chest, waiting for the feeling to return to his legs and wondering if his hair had turned white. Director Iliff and curator Steve McCusker appeared at his side, ashen-faced in the aftermath, and asked if he was all right. Roger nodded

Roger is greeted by herd matriarch Belle after receiving the R. Marlin Perkins Certificate of Excellence for his leadership in developing new ways to manage and care for elephants. (Jim Vincent, *The Oregonian*, 1974)

distractedly, his attention focused on the quiet tableau inside the cage, the matriarch consoling the younger cow and perhaps, in her own way, explaining what happened.

At that moment, Belle turned her head and looked straight at him. Nothing cosmic occurred; there was no inter-species telepathy or otherworldly shared connection, but it was obvious to Roger that they both understood she'd just saved his life. He swallowed hard and nodded, making a silent promise to return the favor if there was ever need. He'd always been fond of Belle, loved her even, and appreciated how their affinity made them an excellent team, but from that moment on he felt bound to her in a way he never expected.

Sadly, it was discovered that Hanako's calf had severe congenital defects. She remained in the nursery for seventeen days and died there.

***

In the spring of 1976, several months after Roger's close brush with death, he was seated at the scarred wooden desk that had been in the barn since Noah built the ark, catching up on the dreaded paperwork, when keeper Jay Haight skidded into the doorway. Haight was relatively new to the zoo, a bright guy by all accounts and something of a wiseacre. Roger didn't really know him and was surprised to see him in the barn, as Haight worked most often as an ethologist—someone who studies animal behavior with an emphasis on the patterns that occur in natural environments. Haight also designed biologically relevant habitats and instructed students in a program run by the Oregon Zoological Research Center.

Haight held out a hand. Clutched in his fingers was a sulcus, the part of an elephant's skull into which a tusk fits. He explained that he'd been walking in the brushy area behind the elephant barn seeking bits of flora to add to a primate exhibit. He was dragging a log and kicking through the undergrowth when the toe of his boot caught on something. When he bent to investigate, he found the sulcus. Roger knew immediately it belonged to Thonglaw. Rather than disrupt the area by searching, he called the police. They painstakingly went over the location where the sulcus was found, but no further evidence emerged. Roger believes that whoever stole Thonglaw's larger tusk either threw it over the fence or slid it under and then buried it in the undergrowth. There it lay, unsuspected by the elephant keepers who passed by a dozen times a day, until the cartilage and subcutaneous tissue rotted enough to release the sulcus from the tusk. When the thief finally retrieved the stolen goods, he never noticed a piece was missing.

The lesser tusk which Roger, with foresight, placed among the hay bales was later displayed in the Lilah Callan Holden Elephant Museum along with a collection of elephant-related artwork and postage stamps, Tamba's retired howdah, a circus elephant's giant red tricycle, and a thirteen-foot mastodon skeleton on loan from the Smithsonian. In January 2013, the museum was dismantled as part of the renovation to the elephant compound. The tusk can now be seen in Forest Hall at the Elephant Lands exhibit at the Oregon Zoo. To this day, the larger tusk has never been found and the thief's identity remains a mystery.

***

In October 1976, Roger and RoseMerrie traveled to St. Louis, Missouri, to attend a joint conference held by the American Association of Zoo Keepers and the American Association of Zoological Parks and Aquariums (now the Association of Zoos and Aquariums). There Roger received the distinguished Edward H. Bean Award, "given in recognition for the most notable birth" of a second-generation Asian elephant. Roger was understandably honored to be the recipient, but all the attention made him uncomfortable. He still felt like a sham, riding on the coattails of greater men like Al Tucker, and his true feelings are probably best expressed by a photograph taken shortly after he returned to Portland. In it, he stands grim-faced and unsmiling, holding the award. Behind him Belle lifts her trunk, mouth open as if shouting the news in celebration. It was a silly, staged moment and he hated it. As he later told friends, the award was nice and he appreciated it, but if it wouldn't help him do his job better, discover a cure for foot rot, or save the elephants, what was the point?

He drew holiday duty on Christmas and brought with him a special morning crew made up of ten-year-old Michelle, seven-year-old Melissa, his dad Leonard and brother Don, and various other male relatives. No one was allowed in with the elephants, of course, but as soon as he'd shifted the animals out of a room, his "Christmas elves" set to work with shovels and brooms to gather up manure, old straw, and any hay leftover from the night feeding.

From his first day at the zoo, Roger regularly shared with his family all but the most harrowing of tales. He didn't want to overly worry anyone, especially RoseMerrie or his mother, and never considered how the job must look to those on the outside until he caught his dad's expression that day as Leonard watched him move among the cows, chaining legs and dispensing their Christmas dinner. In his eyes Roger saw wonderment and admiration. The realization stunned him.

"How can you do it?" Leonard asked. "Go in with them like that? Aren't you scared?"

Roger shrugged and tossed a flake of hay to Tamba. There were times when he was so frightened his bowels shriveled, but he respected the elephants, and they him, and his overall love for them outshone any other emotion. "I always thought you were pretty brave getting in with some of those horses you broke when I was a kid," he said. "They'd have kicked you in the head without a second's thought and left you for dead quicker than these girls."

Leonard nodded and returned to sweeping, his expression thoughtful.

Roger went back to feeding the elephants, a small nugget of warmth inside his chest. For the first time in his life, he felt like maybe his dad didn't find him lacking after all.

A couple of months later, on February 12, 1977, Winkie left the zoo for her new home in Winston, Oregon, having failed to conceive during her ten years in Portland. Despite having a death on her record, she'd never offered a single problem and had remained a quiet and low-key member of the herd. A few days after her departure, Roger brought Me-Tu into the barn for a fetal EKG. An earlier exam revealed Hanako was pregnant again, and they hoped the same for Me-Tu, who'd grown into a roly-poly fifteen-year-old with attitude, at least where her own kind was concerned. A matriarch in the making, she lacked both her mother Rosy's subtlety and Belle's quiet ability to control the herd.

Hanako had exhibited surprisingly good behavior during her EKG, but Me-Tu was clearly "on the prod," as Roger put it, made nervous by the strangers from the local medical school and their equipment. Sullen and brooding, she shifted irritably from side to side with the regularity of a metronome, her head nodding unhappily at being chained on all four legs. It was a temporary measure meant to hold her still for the duration of the test, but Me-Tu didn't understand that. All she knew was weird things were happening to her and she didn't like any of them. Standing beside her head, Roger felt a swell of compassion. They both wanted this nonsense dispensed with as soon as possible.

After the scan was complete and the technicians removed their equipment, Roger bent to release the first leg chain. Much to his surprise, Me-Tu kicked at him. A verbal correction was usually all it took to make her behave, so he growled at her to knock it off.

Hindsight being twenty-twenty, he acknowledges a sane man would have stepped away and given her time to cool down; perhaps brought her a special treat to sweeten her disposition. Instead, he was impatient to get back to work, aggravated by the amount of time the EKG carved from his day, and in no mood to deal with her truculence.

He bent to the chain and Me-Tu kicked again. Roger dodged the blow and had time for a fleeting congratulatory thought—*Oh, Henneous, you agile little fart*—before the leg chain snapped tight, catching him from behind at the level of his belt with such force he became airborne. Body bent in what a bystander later described to Roger as a perfect U, he skated across the floor with his left hand pinned beneath him and came to rest atop a pile of dung and urine-soaked hay.

Beyond the bars of the room, observers stood wide-eyed. One of the vet techs, a woman named Anne Moody, called, "Lay still, Roger!" as someone hurried toward the telephone to summon an ambulance.

Bearded cheek resting against the rough concrete, he blinked at her. "It's cold and wet down here, Anne. Why the hell would I lay still?" He slowly rolled onto his rump and sat up. The back of his left hand was scraped raw, and all that remained of his twenty-dollar wristwatch was the band and rear casing. The rest had been sheared off in his spectacular glissade across the room. Roger held up his wrist for all to see. "Damn her," he said mildly. "She just wrecked my watch."

***

Whenever the zoo needed help with landscaping, sweeping, shoveling, trash pickup, or unloading tractor trailers full of hay, a crew of minimum-security prisoners arrived from the local jail to pitch in. Evaluated on a case-by-case basis, these men could be excluded from the work program based on previous escapes or attempts, sex offender status, assaultive crimes or threats of violence, and negative psychological evaluation. In general, they appreciated the break in routine and the opportunity to be out in the real world, and were well-behaved and orderly, if a trifle attitudinal.

Not ten days after the post-EKG altercation with Me-Tu, a crew from the local jail arrived to assist with spring-prep cleanup of the buildings and grounds. Those assigned to the elephant barn stood slouched under the watchful eye of their guards as they waited to receive their orders, smirking like there wasn't anything new under the sun Roger could show them.

Motioning them forward, he lined them up outside Packy's enclosure, opened the door, and gave them their first glimpse of the magnificent fourteen-year-old bull. Their mouths sagged in disbelief.

Roger flung out one arm like a circus ringmaster. "Behold, gentlemen! That's death on four legs and he don't give a shit about your constitutional rights. If you're foolish enough to go near him, there won't be jack-shit anyone can do except squeegee your sorry ass down the drain when it's over." By the time he finished speaking, every prisoner stood at attention, eyes big as saucers.

That day, a young fellow—a first-time offender—was assigned to be Roger's assistant, and he watched in fascination, maybe even a little envy, as the senior keeper led Rosy into one of the back rooms. When she was comfortably situated, Roger brought in Pet and Tuy Hoa. Last came Me-Tu, who walked beside him with docile obedience until they reached the doorway, where she balked. Her reluctance to enter the room may have stemmed from general pregnancy crankiness or because she was blind on her left side, but Roger guessed she didn't want to be confined with the higher-ranking cows, all of whom she'd challenged repeatedly without success. (The loss of vision occurred while she was on breeding loan to another zoo. Roger never received a solid explanation for the injury and didn't know if it was caused by another elephant, a keeper, or Me-Tu herself.)

Roger remained patient despite the nagging sense of time slipping away, and a liberal application of treats and sweet talk eventually convinced Me-Tu to cross the threshold. Roger checked that all was in order and the cows had plenty of food to occupy them, and headed out of the room. Rather than take the long way around to secure the door behind them, he chose to save a few seconds by passing between Me-Tu and the wall, an opening so narrow he had to turn sideways to clear it.

As he drew even with the shallow cup of her hip, Roger heard a sudden shuffling of feet. He had time to think *Oh, shit,* and begin to squat before Me-Tu was struck hard on her left side by one of the other cows. She and Roger hit the wall together and something snapped inside his left shoulder.

Me-Tu commenced to bawl, submissive language directed at whichever elephant hit her, but also probably at Roger for this terrible breach in etiquette. He slithered out from behind her and shakily exited the room with his left arm hanging limp. Winching the door closed with his uninjured right hand, he sank to his knees on the cold concrete floor, his face sweaty and ashen.

"Roger?" The young inmate's voice creaked like a hinge. "You okay?"

Using his good hand, he felt along the ribs on his left side. Nothing seemed broken and he could breathe without discomfort, but the shoulder worried him. There wasn't much pain, but the whole area felt heavy and useless, almost dead. Roger glanced at the elephants. They were restless, milling about a bit because they understood something bad had happened. Pet looked particularly guilty, so he suspected she was the perpetrator. Me-Tu watched him, tears the size of grapes running down her face. "It's okay, darlin," he assured her, though he knew it wasn't. "It ain't your fault."

From the adjacent room came the sound of jaunty whistling and water spraying from a hose—keeper Bill Wadman going about his business. Roger sent his assistant next door to fetch him. Seconds later, Wadman rounded the corner at a run with an ankus clutched in one hand and the prisoner close on his heels. He slammed to a halt and stared at his boss in dismay. "You're not gonna die on me, are you?"

"No, you asshole," Roger growled. "I think my shoulder blade's broken." He struggled to keep his voice steady and his tone low so as not to further agitate the elephants. The last thing he needed was for the girls to get in an uproar. He told Wadman to order a staff car to take him to the hospital.

Wadman made the call, but Roger's request for an unobtrusive vehicle was superseded the instant it was understood he'd been injured. Like it or not, he rode to the hospital in an ambulance. RoseMerrie appeared in the emergency room a few minutes after his arrival, as did Director Iliff, who repeatedly expressed concern for Roger and his family and offered whatever help they required.

X-rays revealed a lateral split of the left scapula. The break missed the shoulder joint by a mere half-inch, narrowly avoiding an injury that would have required surgery and for Roger

Roger nurses a broken arm at his birthday party, surrounded by friends and family. (Personal photo)

to be trussed up in a "Statue of Liberty" cast for weeks. As it was, he got away with having his left arm and shoulder heavily wrapped and bandaged—his elbow jutting at what passed for a normal angle—and the entire thing anchored to his chest and abdomen with more wrapping. When RoseMerrie heard it would be at least six weeks before her husband returned to work, she rolled her eyes at Iliff. "Thank God I have a job to go to." He laughed so hard he nearly fell down.

To absolutely no one's surprise, Roger was an irritatingly non-compliant patient. He hated being housebound and he chafed at convalescence. At each follow-up appointment, he cajoled the nurses into removing enough of the bandage so they could wash his armpit. When they reapplied the dressing, he'd hold his arm at a slightly lower angle than before. They never caught on, and soon the injured limb was low enough that he could grasp the steering wheel of his truck—never mind that he wasn't supposed to be driving. Three weeks later, he was free of the strapping around his chest and had graduated to a sling he promised to wear, but used only when there were witnesses. By week six, he was back in the barn.

All through his recuperation, his crew kept things running smoothly. "I was blessed with some extraordinary coworkers," he said. "I had people who respected each other. We *loved* each other, though God knows there were times when I could have throttled half of them. I'm not the most sensitive, gentle, tender, discreet individual. I was rarely short on telling them when they weren't cutting it, and woefully short on praising them when they were doing good."

That may be, but his crew's devotion to their cantankerous leader shone through during that difficult time, and would do so again.

Chapter *Eight*

ELEPHANT MOUNTAIN

1978–1979

Roger was at his best in the barn. "Put a shovel in my hand and I'd work all day without complaint," he said. "But paperwork, updating files and such, is my definition of hell." There were reports to fill out, order forms to complete, and elephant files to update. Delaying the job didn't help, because the pile only grew, leaning perilously to one side, ready to slip to the floor if he didn't tend to it.

One afternoon in 1978, he was alone in the office, quietly grousing his way through the stack, lulled by the distant sounds of his crew at work, when there came a knock on the door. Without bothering to look up from the file he was updating, Roger called for whoever it was to come in. The door remained closed.

The knock came again. Roger repeated the invitation, raising his voice slightly, but there was no response. Maybe the noise was the heating system acting up, or one of the ceiling fans. He jotted a note to report it to the maintenance department and went back to work.

Knock-knock.

"Goddammit." Roger threw down his pen, got up, and yanked the door open. No one was there. He checked the alleyway in both directions, but there was no one in sight except Belle, Pet, Tamba, and Me-Tu in the front exhibit room, the four of them contentedly engrossed in a pile of hay.

Roger listened closely, expecting to hear a muffled snort of laughter from one of his crew hidden around the corner, but there was nothing. Maybe he ought to get his hearing checked. Closing the door, he sat down behind the desk and resumed his work.

After several minutes, the knock came again. This time, something a lot stronger than "goddamn" erupted from his mouth. Roger shoved away from the desk and yanked open the door. "If you bunch of jackasses don't have enough to do, I'm happy to—"

The keeper alley was empty.

Belle stood watching him from behind the exhibit room bars. Roger could have sworn she was smiling. "What's up, sweetheart?" he asked.

Bright-eyed, she raised her trunk, curled the tip slightly, and sharply popped it open twice. *Knock-knock.*

Roger laughed until tears squirted from his eyes. Abandoning his paperwork, he slipped between the bars to pay her some attention.

"Belle never failed to get her point across no matter who she was trying to communicate with," he recalled in 2015. "She loved to tease, and was surprisingly adept at using humor to make herself understood. She was more than willing to apply force when it came to the herd, but I can't recall her ever lashing out at one of us."

The same could not be said for Hanako. She'd inherited her mother Tuy Hoa's flighty nature, but now she increasingly displayed bouts of aggression toward her keepers that were reminiscent of her late sire, Thonglaw. She sparred with Me-Tu, seeking to supplant her half-sister in the herd hierarchy, and regularly tormented lower-ranking cows like Tamba and Susi.

Tuy Hoa's labor to produce her irritable daughter had been particularly long and difficult, and Roger wondered whether that might have caused some sort of brain damage in Hanako. Perhaps losing two calves had unhinged her already erratic personality. He'd have given anything to read her mind. The best he could do was warn those who worked with her to have eyes in the back of their head.

The arrival of Hanako's third calf on March 15, 1978 proved as disastrous as Roger feared it would. Born hydrocephalic—a condition in which an abnormal accumulation of cerebrospinal fluid inside the brain causes injury to the tissue—and with a defective heart, the odds were stacked significantly against the little female right from the start. Keepers named her Sumek (Thai for "pretty gray cloud"), but privately called her My-Ow, which means "not wanted," because Hanako shunned this calf as she had her first. Sumek lived out her short allotment of days under the tender care of nursery staff and died five weeks later.

Foot disease continued to plague the barn, and Roger was on the phone nearly every day with one zoo or another seeking advice on how to combat it. Tuy Hoa was his greatest concern. Dr. Schmidt prescribed a combination of antibiotics, anti-inflammatories, and warm foot soaks in hopes of clearing it up, or at least alleviating her discomfort.

Despite a complete lack of interest in Packy, Tuy Hoa continued to be put in with the young bull, spurning his advances as ardently as she'd once embraced Thonglaw's. Packy took rejection poorly and became aggressive in his attempts to force himself on her. When that didn't work, he roughed her up, using his trunk, head, and body to express his frustration. Roger believed these "poundings" exacerbated the degradation of Tuy Hoa's hips. He recommended she be retired as a breeding cow, but he was overruled.

The other cows displayed no such aversion to Packy and the calves kept coming. On May 19, 1978, Me-Tu delivered a handsome, sweet-natured bull the keepers named Khun Chorn, meaning "Mr. Elephant." Just short of one year later, Rosy's son Thongtrii was born, his name a tip of the hat to grandsire Thonglaw.

Roger had just poured a cup of coffee on the morning of June 27, 1979, when the telephone rang. Closing the office door against the noises of the barn, he picked up the receiver. As he listened, the color drained from his face. He dropped into the wooden chair behind the desk as if struck. Pressing the receiver tight against his ear, he listened closely as the Clark County, Washington, police officer on the other end explained that Morgan Berry had been killed by one of his elephants.

Roger was horrified, but not surprised. Given Berry's penchant for working with bulls, he'd been expecting something of this nature for years. The police officer explained that one of Berry's elephants was running loose on the property and he asked if Roger could come and capture it.

Visions of mayhem flashed before Roger's eyes, but they didn't stop him from saying yes, he was on his way. He racked the receiver and gathered a small team comprised of vet tech Anne Moody, veterinarian Michael Schmidt, and a part-time keeper named Doug Groves who had experience working with Berry's elephants and knew them by sight. Loading everyone into his truck along with bull hooks, four sets of leg chains, and a tranquilizer gun, Roger headed north to Kalama, Washington, and the old dairy farm Berry purchased in 1965 and renamed Elephant Mountain.

Roger brooded as he kept pace with morning traffic, his hands tense on the steering wheel. The past year had been a difficult one for Berry. In May 1978, his partner Eloise Berchtold had been killed during a performance in Canada with the Gatini Circus. The guilty elephant was Pet's twelve-year-old son Rajah, the bull Berchtold and Berry purchased from the Oregon Zoo in 1968 and renamed Teak.

Roger had met the vivacious Berchtold a handful of times, but didn't know her well. Glamorous in her spangled costumes, the tall, lithe, determined blonde had trained and performed with a variety of animals including white German shepherds, polar bears, elephants, several species of big cats, horses, and even an aoudad (also known as a Barbary sheep). She met Berry in the late fifties or early sixties, and together they developed several animal acts including "Teak, Thai, and Tunga, the Tuskers of Thailand."

Details of what occurred the night Berchtold died varied depending on the source. Some witnesses recalled Berchtold falling from her perch on Teak's raised right leg. Others said she tripped on a loose piece of matting and fell prone in front of the bull. Teak was variably reported to have stepped on her, flung her across the tent, or plunged his tusks into her. In the pandemonium that ensued, it's not surprising there are so many different versions of the event. Another of Berchtold's elephants, an inoffensive young animal named Thai, burst through the canvas wall of the circus tent. Frightened by the tumult, he fled into the surrounding countryside dragging his leg chain behind him.

Teak refused to leave Berchtold and threatened anyone who attempted to claim her body. In the end, a police sharpshooter fired four .457-caliber rounds into the elephant, killing him. Berry was notified and arrived as quickly as possible. He captured Thai without incident and later donated one of Teak's tusks to be sold at public auction. The proceeds were given to a foundation to rehabilitate prisoners and juvenile delinquents.

News of the tragedy resonated deeply within the circus and zoo communities. Berchtold was well-regarded in the field and liked by everyone, but no one felt her loss more keenly than Berry. By then in his late sixties, he mainly lived alone in the year following her death, spending his hours caring for two wolves and nine elephants, seven of which were feisty young bulls he occasionally set free to wander loose about the property.

Now Berry was gone, too.

Roger and his passengers arrived at Elephant Mountain fifty minutes after leaving Portland. Berry's son Kenny came out of the house to meet them, and said the missing elephant had been spotted once, standing in the middle of the road. An ill-conceived attempt by police to herd it with their patrol car using flashing lights and a siren goaded the animal into charging. At the last instant, it veered off the road and vanished into the thick undergrowth.

Roger cursed and exchanged a sour look with his crew. They listened in growing horror as Kenny related what he'd been able to piece together of his father's final hours.

Berry's health wasn't good, so friends and family kept a close eye on him, particularly since Berchtold's death. Kenny called daily to check in, and Berry telephoned neighbors Joe and Mary Wodeage each night before bed. Last night that call never came, and repeated attempts by the Wodeages to reach Berry were unsuccessful. After notifying the police and Kenny, they went to investigate. They found eight elephants staked and chained as usual, some in the barn and others outdoors, and one missing. Police were understandably reluctant to search the grounds in the dark with an elephant on the loose. Kenny arrived from Seattle two hours later, and a cursory examination of the area around the house and barns produced no sign of his father.

As the sun rose, turning the sky from the milky color of pearl to bright pink and gold, the search area widened and the hunt for the missing man intensified. At last, near a chained outdoor elephant named Buddha, Joe Wodeage and Kenny spied what they thought was an old rag or deer hide. Kenny used a pitchfork to rake it away from the bull and knelt to inspect it more closely, unfolding it piece by piece until he realized it was his father, pressed flat as paper.

Roger turned away, pinching the bridge of his nose to keep from crying. He stared at the forest that surrounded the property and ran a hand over his face. He was already sweating; the day promised to be hot. Glancing across the yard at Buddha, chained between two trees and digging at the ground with his tusks, he asked if they knew for certain Buddha killed Berry.

"We think so," Kenny said miserably. "But it might be the one that's loose."

The words echoed in Roger's head. *The one that's loose.* The one he and his team were meant to find and capture. Sighing, he ordered everyone to gear up with a set of sixteen-foot leg chains carried around their necks to keep their hands free. He, Groves, and Moody armed themselves with bull hooks and Schmidt loaded the tranquilizer gun. Together, they stepped into the woods.

Despite having been logged, the area was dense with undergrowth, the air dim and green where sunshine penetrated the canopy. The forest is a place of natural sounds: wind, water, birds, and other animals. That day, it was silent as a cathedral; the still air weighted with humidity as temperatures climbed into the eighties. Within minutes, they were dripping with sweat. Roger removed his campaign hat, swiped a sleeve across his forehead, and snugged the hat down firmly above his eyes once more.

They headed down the steep hillside, walking abreast in knee-deep logging ruts so as to make less noise and avoid tangling their feet in the undergrowth. Within minutes, the soles of their boots were caked in thick, half-congealed mud they stopped frequently to scrape off. Branches broke beneath them, the sudden noise startling in the eerie quiet.

Roger's nerves jangled. What the hell were they headed into? What sort of elephant were they hunting? Was he friendly? Combative? In musth? The bull was likely frightened by all the goings-on, especially the run-in with police, but where the hell was he? Elephants can walk so quietly as to pass by someone without their knowledge, but in all this tangled underbrush, surely the animal would make some sort of noise.

Roger had a favorite mantra he'd taught his crew: You can *make* an elephant do one of two things: run away or kill you. But you can *get* an elephant to do any number of amazing things. He hoped this particular elephant was willing to listen.

Suddenly, the forest erupted with the unmistakable noise of trees being knocked down. Shrill trumpeting sliced the air, seeming to come from all around. The keepers huddled back to back, trying to spot the bull's location.

When he abruptly popped his head out of an alder thicket several yards away, Groves nearly wept with relief. "Christ, what a blessing!" he said. "That's Thai, and there's not a mean bone in him." Maybe not, but people had died for making that kind of assumption and Roger didn't intend to be one of them. This was, after all, the same animal that escaped the Gatini Circus to run wild the night of Berchtold's death.

Groves, however, chose to bet on the side of the angels. "Thai! Come here!"

The bull trumpeted with delight at hearing a familiar voice and hurried to him. Working quickly, the team ran their chains in a figure-eight pattern around his forelegs to shorten his stride, then attached another length around one rear leg and slowly walked him down the mountain to the main road. Exhausted, thirsty, and hot, Thai collapsed when they got there. An obliging neighbor loaned his garden hose so they could cool down the elephant (and each other). When Thai revived, they got him on his feet, loaded him onto a flatbed offered by a local logger, and ferried him the three miles home.

The details leading up to Morgan Berry's death will never be known, but there are plenty of theories. Some claim Buddha was the killer, and others Thai. Perhaps they're both at fault. The two may have clashed, catching Berry in the middle when he tried to break up the altercation. Perhaps he was laid low by a heart attack while tending Buddha, fell into a vulnerable position, and was trampled.

Guilty or not, Buddha was blamed for the death. Kenny tried for months to place him elsewhere, but not even the Oregon Zoo would take the elephant, despite their excellent track record of managing multiple bulls. Eventually, there was no choice but to euthanize him.

The remainder of Berry's herd was sold. Thai was purchased by the Houston Zoo and lives there still, a venerable old bull in his fifties, feeling his years a bit, but still magnificent. The Oregon Zoo purchased thirteen-year-old Tunga in hopes he would prove to be an enthusiastic breeder and enlarge the gene pool. Another elephant, Ranchipur, was taken by the San Diego Zoo.

Randall Moore, a former employee at Elephant Mountain, took custody of Berry's three adult African elephants. The bull, Tshombe, died, but after much effort and expense Moore was able to reintroduce the two cows, Durga and Owalla, to the African wilderness, where they successfully established their own herd.

Reintroduction of elephants into the wild is an idea growing in popularity. Logging elephants are ideal candidates since they're already accustomed to living and finding food in the forest. The Elephant Reintroduction Foundation maintains three forest sanctuaries in Thailand where over one hundred formerly captive working elephants have been released to live and breed in the wild. As of 2017, over twenty calves have been born.

"While we would all like to see elephants roaming free, the sad truth is that all elephants are managed by humans in some capacity, whether in a zoological facility, in a sanctuary, or in a national park," said April Yoder, chairperson of the Conservation Committee of the Elephant Managers Association.

Two elephant sanctuaries exist in the United States: one in Tennessee, the other in California. Both are privately owned and not open to the public. The elephants are segregated by sex if bulls and cows are present, and no breeding is allowed. According to Todd Montgomery, volunteer and outreach manager at The Elephant Sanctuary in Tennessee, their facility "does not support the breeding of elephants in captivity as there is no indication that these captive-born elephants will ever be part of a viable wild population."

There are those who disagree with that viewpoint, feeling sanctuaries ought to breed elephants as a way to maintain zoo populations that are not self-sustaining. They opine that captive elephants serve as goodwill ambassadors for their wild counterparts by capturing human attention, creating concern for their welfare, and aiding in scientific study that can be applied in the field. For example, DNA extracted from the dung of captive elephants was compared to blood-extracted DNA, and shown to be a reliable source for noninvasive genotyping.

While Montgomery disagrees with breeding, he thinks zoos can provide a positive educational experience for visitors. "The issue at hand is that it is just very difficult to adequately care for elephants in a captive setting in regards to their physical, emotional, and psychological needs, and that includes The Elephant Sanctuary. And some places certainly do a much better job than others."

Those who oppose captive breeding are concerned the animals may wind up living in quarters ill-suited to their needs, displaying behaviors (such as repetitious rocking) that are never seen in the wild. Despite a sanctuary's wide-open areas, such behaviors may still be observed, possibly due to long-ingrained habits. The Elephant Sanctuary in Tennessee reports that their elephants "sometimes exhibit stereotypic behaviors. In some instances, we have seen the frequency of these behaviors decrease over time. We think that the socialization opportunities, enrichment provided from the natural environment, and the privacy of The Sanctuary habitats for the elephants creates an environment that is less conducive to triggering these behaviors."

"Captivity can't serve them as an ark," said former Oregon Zoo veterinarian Mitch Finnegan. "Even if it did, where would you eventually release them to? We've proven that it is much easier to destroy a habitat than to restore it. You could spend all the money in the world and make the best exhibit possible and it is still going to be confinement."

Those who suggest the best way to save elephants is to open the cage doors and return them to the wild make Roger grit his teeth. "Where is this mythical wild?" he said. "When's the last time you saw enough wild to oblige an elephant?"

Linda Reifschneider, president of Asian Elephant Support, agrees. She would like to see land originally belonging to elephants reclaimed and reverted back to wilderness, but securing these tracts is very difficult and many countries will not allow foreign entities to buy such property. Even if one government were to allow it, a change in politics, leaders, or philosophies could complicate, nullify, or undo any steps taken toward conservation.

"In order for elephants to have a serious chance of making it into the next century [and beyond, we need] serious collaboration grounded in reality and science to support range country researchers, veterinarians, mahouts, owners, government personnel concerned with elephant issues, and the people who share where they live with the elephants," said Reifschneider. "Helping them as they work to help the elephants is the best way to make a positive difference. It's the future on which our resources should be focused, not on the deficiencies of the past."

Chapter *Nine*

BUILDING ALLIANCES

1980–1983

Roger knew how fortunate he'd been with regard to partners. First, he inherited Cochran and Robbins from Al Tucker. After they left came Wes Peterson, who'd started on staff the same day as Roger. When Peterson accepted promotion, it was "gentle giant" Gordon Noyes, followed by a slew of others whose devotion to the elephants mirrored Roger's own.

Peterson, in his new role as foreman, arrived one day in the spring of 1980 to inform Roger that his newest partner would be Jay Haight. Roger was less than enthused. Haight had discovered the evidence of Thonglaw's missing tusk and Roger owed him a debt of thanks for partially laying that mystery to rest, but the man's reputation as an independent thinker and a smart-ass worried him. Roger never considered himself the ultimate elephant keeper—there was always room for improvement and he welcomed input—but the last thing he needed in the barn was attitude. He could just imagine Haight swaggering around, scattering opinions like birdseed, doing things his way instead of following rules, and being a jerk in the process. Plus, the man had never worked with elephants. How was Roger supposed to keep the cheeky wiseacre alive while he learned the ropes?

It turned out that his premature assessment of Haight couldn't have been more wrong. The man was opinionated, no doubt about it, but he was also quick-witted, hardworking, thoughtful, profane, conscientious, and innovative, and he possessed a wicked sense of humor. In some respects they were very different individuals, and in others they were the veritable twin sons of different mothers. Once they grew accustomed to each other, Roger and Haight became a perfect team, able to sense each other's moods and read each other's minds.

Haight delighted in springing practical jokes on his new boss. One day, he handed Roger a spray bottle of what appeared to be a well-known brand of automobile body protectant but was actually water mixed with a bit of detergent. He told Roger the latest thing cooked up by the vet was to spray a bit of this on each of Pet's toenails and rub it in with a towel. Willing to try anything when it came to protecting an elephant's feet, Roger fell for it and went to spraying and rubbing away madly. When he discovered he'd been duped, he flung down the towel and exclaimed, "Hell, it made perfect sense to me."

Roger wasn't the only one to whom Haight took a liking. Inexplicably drawn to Hanako, he took it upon himself to ameliorate her aggressive tendencies toward the keepers. With Roger's blessing, he set to work converting the troublesome elephant. First, he brought in Rosy and Tuy Hoa and chained them facing a cement wall. Then he placed Hanako between her mother and auntie and asked her to kneel. He knew from prior observation he had roughly five seconds before she'd try to ram him. At the four-second mark, Haight deftly rolled beneath one of the other cows as Hanako lunged, smacking her head hard against the concrete. She backed up, confused to not find him flattened against the wall, and there he was, smiling at her.

"You want to run that past me again?" he asked, and repeated the order to kneel. Hanako went for him as before and again hit the wall. She was stubborn about it, as he'd known she would be, but he was even more determined. In the end, after several days of this, she capitulated. Something about the way he'd chosen to teach her, or his quiet manner, drew her. Soon, she followed him around like a ten-thousand-pound gray puppy whenever he was in the yard. Her rancorous attitude toward the other keepers didn't fully abate, but she'd do anything for Haight, and he became the one to introduce her to anything new.

Roger was astonished by the transformation and said as much one day as Haight stood by while Hanako had her toenails trimmed. As if on cue, she reached out with her trunk and flipped his campaign hat onto the ground. Roger picked it up, settled it back onto his head, and continued the nail work. Hanako did it again. Once more, Roger put it back on. Hanako flipped it a third time. Without a word, he removed the hat and placed it on the ground beside Haight's boot. For the remainder of the session, Hanako never touched it.

In the winter of 1980, a blizzard howled into Portland, dumping more than twenty inches of snow across the city. Temperatures dipped into the teens at night, rising to the low thirties during the day. Driving to work the morning of the storm, Roger hunched over the truck's steering wheel and squinted past the frenzied swipe of the hardworking wiper blades, eyes peeled for the sudden flash of brake lights ahead or the smear of color as a car lost traction and left the road in a sideways swirl. His tire chains clinked and chimed a muffled rhythm against the snowy macadam.

At the exit for the zoo, a line of orange and white barricades emerged from the blinding rush of snow with a police patrol car parked alongside them, roof lights strobing fitfully in the gloom. A young female officer emerged from the vehicle to explain the road was closed due to ice.

Roger told her he could see that, but he had to get past. "I've got no choice," he said politely. "There's twelve elephants at the top of that hill standing in a night's worth of shit. They're hungry and they're thirsty and I need to check the heat and make sure the damn boiler isn't on the fritz again." He asked her to radio in for permission to let him past, and he listened to the exchange through the partly open door of her cruiser. A voice at the other end interrupted her halfway through her explanation. "Is it the guy with the hat?" In short order, Roger received authorization to advance up the hill.

He turned into the zoo parking lot and gently applied his brakes, knowing there was a layer of ice under the snow but hoping to glide to a halt. His tires stopped, sure enough, but the truck's forward momentum did not. As it sailed across the parking lot, narrowly avoiding light poles, the trunk swung broadside, gaining speed as it skated swiftly toward the plunge above Canyon Road.

Roger clutched the wheel as the truck sped out of control. Any moment now and it would be all over; he'd be dead, or maimed for life. At the last moment, as the truck swept over the curb, its front tires caught on a patch of wind-exposed gravel beneath a small Douglas fir planted at the edge of the tarmac. They bit into the soil and the truck jerked to a halt, rattling Roger's teeth. He threw the gearshift into park and drew a shaky breath. Giving thanks to a deity he claims he doesn't believe in, he left the vehicle where it was, inched his way around the perimeter of the lot, and gingerly headed for the barn.

In February of 1980, eighteen-month-old Khun Chorn was purchased by the Dickerson Park Zoo in Springfield, Missouri. Pending the arrival of their truck, Roger placed the calf in quarantine. The affable youngster, a favorite with everyone, was

Khun-Chorn at Dickerson Park Zoo. (Photo courtesy of Melinda Arnold)

lonesome in his isolation, so keepers made it a point to visit daily and spend as much time with him as possible. His pen was floored with woodchips which he delighted in throwing all over the room and himself. Every day someone would clean him off and provide a good scratch with a narrow, long-handled brush.

"Koonie," as he was affectionately called, already understood leg chains because Roger made it a point to introduce calves to the restraint device as early as possible. "They'd try to run away and they'd hit the end of the chain and dump themselves and have a screaming temper tantrum," he said in 2015 when questioned about the use of chains. "But they were always right where mom could check on them and see what the big deal was and she knew they weren't hurt." Once the initial struggle was over and the calf realized it wasn't being murdered, the procedure became a routine to be enjoyed, particularly the part involving a tasty reward.

Roger's elephants were chained every day, never for very long, and were always rewarded for their compliance. The chains were used primarily as a training tool, to hold them while they received medical care or blood draws, to keep a curious animal from wandering into trouble, and during mealtimes to prevent higher-ranking elephants from poaching food. Stories of elephants being confined to one spot, often in isolation, for days, weeks, months, and even years, made Roger nearly incoherent with rage. Over twenty years later, he still has the same response to the idea. "Leaving an elephant chained long-term is a cop-out on the part of the trainer. There's no good reason for it. You'll only hurt the animal."

One day, he visited Khun Chorn to perform routine foot maintenance. When he was finished, he tapped the foot on the tub and said, "Koonie, foot," the command to lower it to the floor. What he didn't understand was the elephant was inadvertently standing on the chain for that foot and couldn't lift it to follow the command.

When he didn't respond as expected, Roger tapped the foot again and repeated, "Koonie, foot." In an effort to make this idiot human comprehend his predicament, the calf reared back and balanced on his hind legs. Roger nearly went over backward in surprise and praised him effusively with treats, rubs, and scratches.

Warren Iliff made Roger a promise years earlier to press ahead with the request for an elephant restraint device in the barn. Since then, he'd worked diligently to sell the idea to those with their hands on the facility's purse strings, citing its necessity in caring for certain difficult animals, reminding them of the tragedy of Thonglaw and the threat that the same could happen to Packy. There were two breeding bulls in the barn now, but Tunga seemed disinclined to procreate. That left the job to Packy.

Not above laying on guilt when necessary, Iliff pointed out that if anything bad happened to the zoo's pride and joy, particularly something avoidable, the citizens of Portland would not only mourn, they'd arrive at the gates with pitchforks and torches. The breeding program that had elevated the zoo's status and brought in thousands of visitors (and their dollars) could become a thing of the past.

His persuasion and persistence paid off. The elephant barn underwent extensive renovation, including the addition of separate hay and feed storage, a second yard, a manure dump, public restrooms and, at long last, a $200,000 elephant restraint device, a joint creative design effort between Roger and engineer Jim Riccio.

Now they could finally provide all the elephants with the same level of care, and Roger set about training the elephants to accept the device. He began with Belle and Rosy because where the matriarchs went, so went the herd. Because he'd asked that the ERD be built to serve as a pass-through to the outdoors, the elephants quickly learned to ignore the odd new contraption and enter it willingly, knowing each time they went through there would be a reward of carrots, bananas, or hay at the other end.

Almost immediately, every zoo across the country took notice. Portland's ERD was the first of its kind in the United States and Roger's phone rang off the hook with questions about its cost, construction, and use. When one of his callers referred to him as a world-class authority on the device, he scoffed. "If I can help somebody and maybe keep them alive, I'm morally obligated, but as for being a world-class authority?

Hell, I'm not even a world-class authority on me!" Soon, every zoo in the nation was budgeting for an ERD of their own.

Roger stressed to every caller that the ERD wasn't magic, nor was it a replacement for well-trained animals and knowledgeable staff. The "why" behind its use was at least as important as the "how." If they used the ERD as punishment, it wouldn't be long before the elephants refused to move through it.

That would have been Packy's choice from the start. Always reluctant to embrace change, he took the longest to acclimate to the ERD. Once he was out in the new yard, however, he wanted the world to know his displeasure and he took to charging the backyard feed wall, reaching up with his trunk to try and grab a keeper. When Roger heard about it, he wasted no time in delivering an attitude adjustment. Standing on the wall in clear view as the bull charged, Roger waited until Packy was in range, then bellowed, "Packy, no!" at the top of his lungs and nailed him at the base of the trunk with a shot from a pellet gun that caused no injury, but stung like hell. Packy slid to a halt, backed up, shook his head, and walked away. He never attacked the wall again.

With the advancement of the ERD, a new way of working with elephants emerged. "Protected contact" promotes both animal and keeper safety by no longer requiring handlers to enter cages. With the help of ample rewards, elephants are trained to willingly enter the device to receive treatment and learn to present a specific body part (such as their feet or ears) to a particular opening for examination. Roger understood the theory behind protected contact, but disagreed with using it exclusively. Hands-on work could certainly be done employing the ERD, but he believed the elephants ought to be familiar with keepers entering and leaving their enclosure. "God forbid you have a medical crisis and need to go in with an animal that has no experience of you being in its space," he said in 2017.

Emerging science affected the zoo in many different ways, but in the early days there was nothing behind the elephant breeding program but conjecture, hope, and an enthusiastic bull named Thonglaw. All that was understood about elephant reproduction was the general notion of egg-meets-sperm. The few articles published on the topic suggested estrus was a seasonal occurrence lasting approximately twenty-two days, but no one knew for certain, in part because elephant cows (unlike most female mammals) do not engage in sexual invitational displays to signal their readiness to mate. Despite the continued success of the breeding program (since its inception in 1962, nineteen pregnancies had resulted in fourteen calves surviving past their first birthday), it remained an inexact science.

Shortly after his arrival at the zoo in the 1970s, veterinarian Michael Schmidt initiated pioneering research into mapping the elephants' estrus cycle with an eye toward cracking the code to artificial insemination (AI). Establishing estrus was just the first step. Reliable methods for collecting semen, freezing and thawing sperm, and performing the insemination needed to be developed, as well. If that could be accomplished, elephant cows would be spared the disruption of herd dynamics and the stress of being transported for breeding, sometimes over great distances. They could remain in their home zoos and the sperm would come to them.

Since the advent of his research, Schmidt and his team of scientists and keepers determined a healthy cow ovulated every sixteen weeks or so (roughly three times a year; much less frequently than most mammals), but the egg was viable for only 48–72 hours. The next goal was to discover a reliable means to pinpoint that window of opportunity.

Elizabeth "Bets" Rasmussen was a wiry research professor at Oregon Health and Science University when she first observed Packy dabbing the end of his trunk onto a damp patch of sandy soil and curling it against the roof of his mouth. His repeated performance of this action—called a flehmen response—and the bull's highly visible erection convinced her of the presence of a sex pheromone. She believed if she could identify and isolate the chemical compound that aroused Packy, there was an excellent chance of more fully establishing the estrus cycle in elephants. Although the substance in question was one she could neither see, smell, nor adequately define, Rasmussen relished the challenge. Excited by the possibilities, she immediately set aside her current research. The very next day, in collaboration with Schmidt, she embarked on a field of study lasting fifteen years.

Rasmussen, being very down-to-earth, developed a wonderful rapport with Roger and his keepers. At her request, they collected elephant urine daily, eventually amassing close to 7,000 gallons she stored in various refrigerators and freezers all over Portland, transporting the containers of urine in an old Mustang with the rear seat taken out.

"You could always smell her coming," said Roger.

He once asked her what she would do with the car once the research was over. "I'm going to have it bronzed," she replied.

Rasmussen's research proved successful, and she was able to better pinpoint the time frame in which the egg is most viable. These days, endocrine research and regular monitoring of hormone levels are two of the tools used to determine a cow's readiness to mate. Once conception occurs, transabdominal ultrasound is employed to oversee the development of the fetus and monitor the birth process.

Unfortunately for Dr. Schmidt, his dream of achieving the first successful artificial insemination in an elephant went unrealized. That honor fell to Dr. Dennis Schmitt of Dickerson Park Zoo in Springfield, Missouri, with the arrival of a calf named Haji in 1999. As of 2016, about twenty AI calves have been delivered.

***

As the battle against foot rot rolled on into 1981, Tuy Hoa continued to be Roger's primary focus. The arthritis in her hips had advanced to such a degree that the twenty-six-year-old, who should have been in her prime, could no longer straddle like a normal elephant when urinating. Her feet and legs remained wet, exacerbating her chronic foot issues and causing infections leading to loss of tissue integrity.

Roger's need for good work boots led him to Bill Danner and his shoe manufacturing company, and the two men fell into an easy friendship. When Danner heard about Tuy Hoa's situation, he offered to have his craftsmen build a pair of boots to keep her back feet clean and dry. The company constructed a pair of leather elephant-sized hikers complete with metal eyelets, laces, and even the Danner logo tag. After sniffing them over thoroughly, Tuy Hoa stood placidly while Roger tugged them onto her feet. She seemed to understand they were meant to help and never tried to take them off.

On November 13, 1981, four-year-old Thongtrii was sold to the Roeding Park Zoo (now the Fresno Chaffee Zoo) in Fresno, California. Roger never liked seeing one of his kids leave, but he felt confident the little elephant would be well taken care of. He'd met the director of the zoo, Dr. Paul Chaffee, when Chaffee visited Portland for a discussion on hiring and retaining employees. When Chafee asked Roger for a description of his job, Roger replied, "Days, weeks, and months of backbreaking labor punctuated by moments of abject terror."

As the conversation moved around the table, most recommendations involved employee incentives like improved health benefits, increased salary, and vacations. When it came Roger's turn to comment, he leaned forward in his chair, cocked back the brim of his hat, and fixed a steely-eyed gaze on the zoo's esteemed guest.

"How would I motivate an employee? First, I'd try to teach by example, help them learn the routine and show them how it's done. If that didn't work, I'd drive them with a whip. And if I couldn't drive them, I'd send their sorry ass down the road and find someone who *wanted* the job!"

Chaffee roared with laughter. As a man who'd fought diligently to initiate new programs in nutrition, quarantine, treatment, and education in order to meet the growing professional standards of zoos all across the country, he appreciated a no-bullshit attitude. The two men remained friendly, and Roger felt certain Thongtrii would be spoiled rotten in his new home.

In October 1982, Hanako delivered what would be her last calf, an attractive little bull the keepers named Look Chai. To the relief of all, he proved to be a robust and healthy little

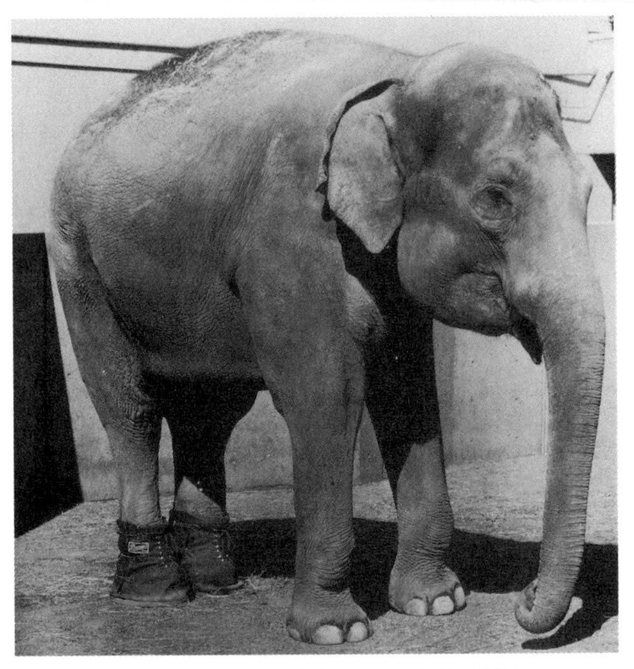

Tuy Hoa wearing two custom-made boots provided by Bill Danner to keep her feet dry and encourage healing. (Photo courtesy of Danner Boots)

fellow and was soon happily dashing about the exhibit room, weaving between the legs of his mother and whichever of his aunties was present. Belle, of course, was particularly taken with the newcomer and stole him whenever possible.

The day after Christmas, Roger arrived in the barn to find senior primate keeper Dave Thomas gasping for breath. He'd dropped by to see Roger and discovered Pet's newborn daughter standing in the keeper alley, having slipped easily between the bars of the exhibit room. For the past hour, she'd led him in a merry chase around the back area while her mother and aunties watched with undisguised amusement. Laughing, Roger helped corral the energetic little miss and put in an order for chains to be strung between the bars to keep her—and any future calves—out of mischief. His standard greeting of "Good morning, sunshine!" prompted Haight to christen her Sung-Surin, Thai for "sunshine."

Two-month-old Look Chai was delighted by his new playmate and often lay down on the ground to encourage her to climb on him. When she tired of that game, he'd run ahead, taunting her to give chase, then spin around and pursue her.

Not quite three months later, on April Fools' Day 1983, Rosy delivered Rama, whose name means "pleasing" in Sanskrit and "high and exalted" in Hebrew. The three elephant children became inseparable, squealing with delight as they chased each other, ears flapping, tails rigid, and uncontrollable little trunks flailing the air. They reminded Roger so much of a litter of piglets that he nicknamed them "the pig pile."

Rosy had been taken from her own clan at a perilously young age, but she'd skillfully created a family herd within the Oregon Zoo. Although she'd never had the early experiences granted to calves growing up in a herd—the sounds, smells, and circumstances that would inform her future—her personality allowed her to roll with what she had, learning as she went. She'd successfully delivered more calves than any other cow in the barn. By the time Rama arrived, she was an expert.

As part of an ongoing research project at the time, newborn Rama was placed under twenty-four-hour surveillance. Roger's daughter Michelle was drafted onto the research team because she could tell the elephants apart by sight. Now sixteen years old, Michelle had been volunteering at the zoo for more than four years. After training for ten weeks at the Children's Zoo, where she cared for animals such as ducks, rabbits, and goats, she proved responsible and trustworthy enough to volunteer at the main zoo.

Michelle shared her dad's enthusiasm for elephants, but an allergy to the timothy hay they were fed kept her out of the barn except on special occasions, like when a calf was born. Roger needed to find her a different location to work, but being a protective father, he had a firm sense of who was—and was not—suitable company for his impressionable teenage daughter. He chose soft-spoken senior primate keeper Dave Thomas, a man he much admired. Michelle so enjoyed being with the apes and learning from Thomas that she volunteered over summer vacations and every Sunday and holiday, walking over to the elephant barn to share lunch and a thermos of coffee with her dad.

Now she was back in the elephant barn, seated on the floor against the glass between the visitor viewing area and the keeper alley, armed with a clipboard, an interval timer, a pen, and a box of tissue to combat her runny nose. Something of a night owl, she enjoyed working the quieter midnight shift.

As dawn approached, she noticed something amiss. When Rama straddled to pee, Michelle spotted what looked like blood in his urine. She reported it to Roger when he arrived, and he went into the cage to investigate. It turned out Rama had a raging urinary tract infection, and Roger knew exactly how it happened.

He often joked that the goal in the barn was to "get the turd before it hits the floor." The barn crew expended immense effort in trying to keep the floors as clean, hosed, and dry as possible to prevent the elephants from tracking through their own waste and increasing the risk of foot infection. Nowhere was the pursuit of cleanliness more pronounced than in a maternity ward. Unfortunately, Rosy defecated immediately following Rama's birth. Before anyone could get inside the cage with a shovel to clear it away, the wobbly newborn took a tumble and skated through the excrement on his belly, exposing his raw umbilicus to bacteria.

The veterinarians prescribed a course of penicillin to be dispensed by needle twice a day for one week. While Rama received his treatment, Rosy would be temporarily secured at the far end of the room by a single rear leg chain to prevent her from interfering.

The serum was glutinous and took several minutes to drain from the syringe. It had to have hurt some, pressing in beneath his hide, but Rama stood quietly on that first day and accepted the shot without complaint. By the second day, however, he'd wised up. He allowed the keepers to herd him to the far end of the cage without argument, but before the needle even touched him, he began to struggle. As Roger and his team fought to restrain more than two hundred pounds of wriggling baby elephant, the little bull began to scream.

Rosy was engrossed in a pile of hay at the other end of the room when the first cry rang out. Her head shot up in alarm, ears fanned out, eyes bulging. Rather than lunge forward, fighting her restraint, she stepped backward onto it. The connecting link that held the chain to the floor snapped with a report as sharp as gunshot.

Roger knew immediately what the sound meant and bellowed, "Everybody out!" The keepers scrambled to safety beyond the cage bars, but he was sandwiched between Rama and the wall with nowhere to run. Trapped, he watched as more than eight thousand pounds of pissed-off mother pachyderm hurtled toward him. As had happened with Hanako, time slowed to a torpid crawl. He had a fleeting moment in which to mourn for RoseMerrie and their girls.

To his astonishment, Rosy slid to a halt just inches away. Towering over him like an avenging goddess, she bestowed a look of withering contempt on the cowering man, wrapped her trunk around Rama, and marched him back to the far side of the room where she resumed eating as if nothing happened.

Outside the cage, Haight laughed weakly. "I believe she just said, 'Pardon me, you jerks, but this belongs to me.'"

No one appreciated a quip more than Roger, but at the moment it was difficult to work up a mouthful of spit, let alone a sense of humor. He shakily slipped between the bars to stand with his crew. He wanted a cigarette in the worst way ("A beer would have gone down a treat too," he later recalled), but there was a bigger issue to settle than his rattled nerves. If they meant to administer Rama's medication twice a day, they needed Rosy to cooperate.

Some keepers would have beaten her for what she'd done, but Roger hadn't been trained that way, nor was it in his disposition to punish an elephant with physical abuse. He understood Rosy was driven by instinct, not malice, and he couldn't discipline her for that. He watched her lift hay to her mouth, Rama safely tucked between her front legs; it was a scene of such peace and contentment, you'd never have known a crisis nearly occurred. Credit for averting it went to Rosy. God knows he hadn't done anything except stand there waiting to die. And that was the thing. Some elephants *would* have killed him, but Rosy chose not to. That spoke volumes about how much she understood what the men were trying to do. Maybe she didn't oppose Rama's medical treatment so much as the keeper's chosen technique.

Roger decided to trust her. He explained to his crew that they were going to chain Rosy again, but this time in a corner with her rump to the wall and a man on each side. Then they would place "that little bonehead" between her front legs and treat him right there.

"You're crazy," someone said. "She nearly killed you."

"Actually, she didn't. She never even came close." He waited a moment while that sank in. The plan wasn't perfect, he acknowledged, and if someone had a better suggestion, he was ready to hear it. When no one offered an alternative, he nodded and started toward the elephants. "Look at it this way," he told the men at his heels. "It'll either work like a charm and they'll write songs about it, or I'm gonna die and it won't matter."

Rosy accepted the leg chain without argument, rumbling softly in her throat as the men secured it to the floor. Her trunk caressed a sleeve, a pant leg, a shoulder as the men worked to position Rama. With everything going on right where she could observe closely, she readily allowed them to inject her whiny offspring. The plan worked so well they repeated it twice a day for the rest of the week. Each time, Rosy behaved with calm deportment, although the half-watermelon she received after each successful session may have had something to do with it. Squeaking with pleasure, she'd crush the fruit between her massive molars, raining juice down on Roger and her calf.

Rama's mischievous streak emerged early. One afternoon as Rosy, Tuy Hoa, and Hanako were being chained prior to receiving lunch, he decided to misbehave. Flitting around the enclosure with his ears fanned wide and his trunk and tail aloft, the youngster initiated a game of tag with his keepers. Dodging this way and that, he uttered elephantine "nyah-nyah-nyahs" whenever the men missed a grab.

The hungry cows weren't stupid. They understood what, or rather *who*, was impeding delivery of their meal and it didn't take long for Rosy to lend a trunk in disciplining her wayward offspring. The next time Rama came within reach, she slapped the snot out of him. Reeling from the blow, he hastened to her side and waited meekly for his own chain to be attached. From then on, whenever he was asked to present his foot, he'd glance at Rosy and then offer it without argument.

Tuy Hoa was only twenty-eight, but in many ways her body was already ancient. Despite diligent care, arthritis continued to cause her distress. The boots provided by Bill Danner helped over the short term, but nothing seemed to conquer the foot rot. Roger lay awake at night wracking his brain, desperate to come up with a solution to her suffering. The usual methods provided only temporary relief at best. Watching her shuffle from place to place, moving as little as possible, he knew it was only a matter of time.

On April 20, 1983, veterinarians made the decision to euthanize Tuy Hoa. As they had with Thonglaw, the barn crew came together to see her out, then donned protective clothing and disposed of her corpse over a span of hours, working grimly with their saws, speaking little and quietly going their separate ways as soon as possible to mourn as they chose. That evening at home, Roger sat alone on his back stoop, drank cheap rye, and wept.

Tuy Hoa's death affected more than just her keepers; it left Rosy without her lifelong companion and Hanako without her mother. Herd dynamics shifted and swirled as the pecking order changed. Both Hanako and Me-Tu grew increasingly combative, eager to raise their standing in the herd. Together, they took on Tamba and Susi most often, but on occasion were brave enough to challenge Rosy and Belle, only to suffer defeat. In an effort to calm things down, Roger rearranged the two herds, separating Me-Tu and Hanako, hoping that without the other to egg her on, each would become less aggressive. Instead, Hanako took out her grief on Belle by challenging her for the matriarchy. Belle bested the younger cow, but the elbow of her right front leg was injured. Vets treated the pain with oral and injected anti-inflammatories, narcotics, and topical DMSO (dimethylsulfoxide) to decrease pain. Even so, she refused to cooperate with physical therapy no matter what Roger did to win her compliance.

As the summer of 1983 passed, Belle grew increasingly lame, unwilling to use her injured elbow. The balance of her weight shifted, putting increased pressure on her left side as she struggled to alleviate discomfort on her right. Slowly and steadily, the elbow's range of motion grew smaller and smaller as summer gave way to fall, and fall to winter.

The weekend before Thanksgiving, the "Master of Disaster" arrived. Wild-born Hugo had spent his life as a performer, most recently with Ringling Bros. The twenty-three-year-old had developed into what's referred to as a "junkyard elephant," an animal so aggressive no one wants it. Ringling planned to euthanize him, but someone suggested offering him to the Oregon Zoo as a breeding bull in exchange for the first calf he sired, assuming it was healthy and Ringling wanted it. The zoo agreed.

"Hugo wasn't handsome, but he was highly intelligent and death on four legs," Roger recalled in 2015. "Your worst nightmare amplified. He'd respect you, but only if you showed him you weren't going to macho him all the time." During his first six months at the zoo, the bull gained a thousand pounds and likely thought he'd died and gone to heaven.

Because Hugo understood the principles of leverage, he nearly destroyed every door in the barn. Placing the broad base of his trunk against a door, he'd use his hind legs to drive off from the floor and flex the entire thing until it nearly came off its track. These were not flimsy doors by any means, but four-inch-thick, high-tensile concrete with rebar placed every foot in a checkerboard pattern. He never managed to shatter them, but he cracked each one, forcing the zoo to replace them with six-inch doors reinforced with eight-inch rebar. Those he couldn't flex, but that's not to say he didn't try.

Chapter *Ten*

DODGING DISASTER

1984–1987

Gunther Gebel-Williams was talented and charismatic, a media star, the most celebrated circus performer of his generation, and widely regarded as one of the greatest animal trainers of all time. With his surfer-blond hair, gleaming smile, muscular form, and dazzling, sequined outfits, he captured the imaginations of those who saw him perform in person or on television. Roger watched from his living room countless times, transfixed by the image of this Goliath simultaneously putting as many as seventeen tigers through their paces. What most astonished him was Gebel-Williams's ability to work elephants and tigers together, convincing the elephants to allow the big cats, their natural enemies, to ride on their backs.

When he learned Gebel-Williams was coming to the Oregon Zoo in 1984 to observe Hanako's son, Look Chai, and consider adding him to his act for Ringling Bros., Roger felt almost giddy at the prospect. His attitude about making elephants perform hadn't changed from his early days, but he admired Gebel-Williams's ability to work well with many different species by using training methods founded on respect, just as Roger did. The two men hit it off immediately, united by their shared view of animals as dependable and honest.

Gebel-Williams was delighted with Look Chai, who'd grown into a handsome two-year-old with tushes roughly two inches in diameter and about eight inches long. Gebel-Williams agreed to purchase him, and the next day Roger and Haight moved the calf into quarantine to await the arrival of a Ringling truck. They made him comfortable, securing him with a single leg chain to a bolt threaded through the cinder block wall and left him there while they fetched Rama, who was being temporarily moved to Point Defiance Zoo in Tacoma. Roger was saddened to see the pig pile go their separate ways, but with fourteen animals in the herd, five of them bulls, the barn was at capacity and room had to be made. Sung-Surin remained, at least, and that was some consolation.

Unhappy at being separated from his herd and left alone, Look Chai pulled his leg chain anchor out of the wall while the men were gone and tangled himself in the links. When they returned with Rama, Roger attempted to free him, but the "seriously pissed off" six-hundred-pounder attacked and tried to skewer Roger's foot with his tushes.

First Hanako, and now her son. Roger, who had set aside his ankus to assist the calf, fought back, yelling and swearing. Haight managed to move Look Chai away and calm him down. They rechained him—this time with a bolt that went *through* the wall, not just into it—and left the two calves to commiserate.

Gebel-Williams found Look Chai to be a gentle, intelligent, and willing performer. Having forged a lifelong friendship with Roger, the trainer made a point of keeping him up-to-date on the bull's progress. During one performance, Look Chai (now called Sabu) worked beautifully—completing hind-leg stands, sit-ups, and front-leg stands as requested—eager for the treats his trainer surreptitiously palmed into his mouth after every successful completion. Unbeknownst to Gebel-Williams, one of Look Chai's tushes had somehow become damaged and now sat loose in its sulcus. As they trotted together around the circumference of the circus ring, Gebel-Williams waving to the audience, Sabu reached up with his trunk and plucked the broken tush out of its socket where it was rattling around. Much like a child handing something to its parent to carry, he calmly offered the tush to his trainer and, like a true show business professional, carried on without missing a single step.

In his own way, Roger did the same. He arrived at work no later than seven, spent the day with the animals he loved best in the world, and returned home in time for dinner. Gone were the days of filling the cookie jars with snickerdoodles and reading to his girls before bed. Michelle was having a tough time at college, and Melissa would soon graduate from high school and leave the nest as well. He couldn't quite believe how quickly it all happened. One minute they were babies in

"Look Chai" or Sabu, roaming the PAWS sanctuary in San Andreas, CA. (Photo courtesy of Kim Gardner and PAWSweb.org)

diapers and the next they were asking to borrow the car. He'd tried hard to be a caring and responsible dad; he hadn't missed a single parent-teacher conference or school performance, but he hadn't been as patient as maybe he ought to have been. He felt certain he'd neglected them and RoseMerrie because of all the extra hours he spent in the barn. They'd made it clear they understood he was doing important work, and they were proud of his devotion to the elephants, but even their reassurance didn't assuage his nagging sense of guilt.

RoseMerrie was working full-time as a secretary for a local school district, a job she loved, when she discovered a lump in her breast in 1984. Rather than panic, she did the prudent thing and immediately went to the doctor. The cancer diagnosis sent Roger reeling with fear, terrified that he was about to lose the woman he loved, the perfect wife, the one he'd never quite felt he deserved. The small tumor required no surgery and was easily treated with radiation. RoseMerrie and her doctor were confident of a full recovery, but Roger was a wreck, distracted and preoccupied. He asked his crew to double-check every order he issued, and told them if anything seemed senseless or stupid or unsafe, they were to grab him by the ears, look him in the eye, and shout, "Is anybody home in there?" They laughed, but they kept a watchful eye on their boss.

Me-Tu may have looked like her mother Rosy, but she was every inch Thonglaw's daughter; pugnacious and confident, almost swaggering in manner. Longtime Portlanders who recalled her birth sixth months after Packy's thought she had a sibling rivalry with her older half-brother. They assumed she was resentful of his celebrity status as first-born, an honor she felt should have been hers.

She was in the front-room exhibit area with Rosy, Hanako, and Tamba when Roger received an order from Dr. Schmidt to draw a routine blood sample from the young, pregnant cow.

Roger asked Haight to assist, and together they moved Me-Tu into a smaller adjacent area, a quieter room away from other animals and the noise of visitors. Blood draws were common practice, but not always easy to perform. Veins are most prominent on the backs of an elephant's ears, a particularly sensitive area, which can make venipuncture difficult.

The men and Me-Tu were old hands at the procedure and Roger expected things to go smoothly. He stood by her head, palming carrots into her mouth, murmuring soft words of assurance that it would all be over in a minute or two provided she was a good girl. Meanwhile, Haight swabbed the area behind her ear with alcohol and picked up the needle.

It could have been that Roger was preoccupied with concern over RoseMerrie and not as attuned as usual, or that Haight's needle-stick hit a particularly delicate area. Perhaps Me-Tu realized what was coming and decided to vent her displeasure. In any case, she was suddenly in motion. Rolling her trunk up to her forehead, she shoved the broad underside hard against Haight's belt buckle and ran him straight up the wall as high as she could reach, pinning him to the concrete like a butterfly in a display case.

Roger caught the corner of her mouth with the hooked end of his ankus and yanked hard, yelling for her to get back. He stared helplessly at his partner suspended high overhead and thought, *I'm watching a dead man.*

Not so. Haight was very much alive and his agile mind, both a delight and an aggravation to his boss over the years, was galloping for all it was worth. He knew Me-Tu would quickly come to her senses and release him. When she did, he would drop to the concrete directly in front of her, in a perfect position to be trampled. He had seconds to decide on a plan of action.

Me-Tu realized she'd transgressed in the worst possible way. Squeaking her apologies, she stepped back, releasing Haight. In a move to rival the best circus acrobat, he kicked off from the wall as hard as he could and, with incredible luck, landed in an empty water tank. He crouched on the bottom, making himself as small a target as possible, expecting the elephant to come after him at any moment. When she didn't, he gingerly peered over the rim of the tank.

"Jay!" Roger's voice was an explosion of relief.

"I'm okay," Haight said weakly, though he knew he wasn't. Every breath was agony, which meant he had at least one broken rib. Grimacing, he pressed a hand to his side, slowly climbed from the tank, and staggered out of the room to find the nearest chair.

Despite Roger's insistence they call an ambulance, Haight drove himself to the hospital, where doctors in the emergency room found three cracked ribs and some ruptured blood vessels in his eyes. Haight had gotten off easy and he knew it.

Roger's relief was profound. One of his chief concerns was to end each day with everyone, keepers as well as elephants, safe and intact. His biggest fear was of making a mistake that got someone hurt or killed.

In those days of free contact, elephants accounted for more handler injuries than any other large animal chiefly because keepers tended to work directly with them. Despite the trappings of domesticity, "they're wild animals and as such don't necessarily love you, and you'd best not forget that," said Roger in 2016. "As majestic, intelligent, and intuitive as they are, elephants can be dangerous, pure and simple, and you should never take their behavior for granted." In 1994, the Department of Labor Statistics listed "elephant handler" as one of the riskiest jobs in the country, ahead of even firefighters and police.

Dealing with elephants in the confines of their enclosures was nothing to take lightly, but the thought of having one break and run—as Tamba had done with the inexperienced handler—made Roger's blood run cold. So when he received a call from one of the Point Defiance Zoo elephant keepers informing him they were sending Rama home to Portland after the elephant had escaped from his handler, Roger expected the worst. But the man on the phone couldn't stop laughing, which Roger figured was a good sign. At least it meant nobody had died. By the time the man finished his story, Roger was chuckling, too.

When Rama arrived in Tacoma in 1984, the plan was for him to remain there until he began to show signs of musth, at which point he'd return to Portland and be put to stud. The one-year-old elephant was personable and attractive, and the crowds loved him. He was so easy to work with that keepers took to walking him through the grounds early in the morning before the gates opened and kept to the habit, even as Rama grew into a handsome six-year-old.

A couple of days earlier, a tractor trailer backfired as Rama walked with his handler through the parking lot on his daily constitutional. When the startled elephant shot ahead a few steps, the handler grabbed his tail to keep him from getting away. Convinced that whatever had made the loud noise now had him firmly by the nether regions, Rama panicked and bolted into the surrounding neighborhood, tearing down trees and bushes as he ran. The astonished keeper raced after him.

A mile away, a woman washing dishes at her kitchen sink, with the window above open to let in the warm breeze, experienced the shock of her life when her fence suddenly exploded and Rama hurtled into her yard. He stuck his trunk through the open window and into the sink, found the sudsy water not to his liking, and busted through another section of fence just as his breathless handler arrived.

After a prolonged roundabout chase, they wound up in the parking lot of a Piggly Wiggly grocery store. Rama strolled up and down the rows of parked cars, sniffing bumpers. The handler asked someone to call the zoo while he kept an eye on the elephant, and it wasn't long before the curator arrived. He told the handler to buy some bananas to keep Rama occupied until the vet arrived with backup, but the handler said he didn't have any money. No doubt worried that Rama might spontaneously take off again, the curator yelled at the handler to steal them if he had to.

There was no need for theft. The Piggly Wiggly manager obligingly donated a case of bananas to the cause while customers remained safely indoors, watching through the windows. Eating kept Rama occupied until reinforcements arrived, and the crowd erupted into cheers and applause as he stepped into the truck for the return trip to Point Defiance.

***

Medical and cognitive testing had been a routine part of life in the elephant barn for decades despite protests from Morgan Berry and veterinarian Matt Maberry, who believed that the use of zoo animals for experimentation or private research was unethical. In the late 60s, Pet and three other elephants had been involved in a study to determine visual acuity, dexterity, and intelligence. The test consisted of a slide projector and a custom-built box with a screen on either side of which was a large white button. The elephant was supposed to push the right-hand button when shown a white slide, and the left-hand button when shown a barred slide. The images were disclosed at random, with no discernible pattern the elephant might memorize. For each correct response, a sugar cube was delivered down a tube. Once twenty correct responses had been recorded, the trial ended for the day. Although some of the elephants struggled at first to master the test, all eventually succeeded.

The researchers returned in 1986 to test the elephants again and see if they remembered the study. Three of them scored twenty correct responses almost immediately, but Pet labored.

She'd get twelve right and miss one, then she'd get fourteen right and make an error, never quite reaching the desired goal. The scientists looked on, one callously remarking that Pet was really stupid compared to the other elephants.

Standing nearby, Haight took issue with the comment and suggested the researcher look at it from the elephant's perspective. Pet had clearly figured out that twenty correct answers brought an end to the sugar cubes. Playing the game her way, she scored far more treats than her supposedly smarter classmates. The scientist realized he'd been superbly played and Pet received an extra sugar cube for being so smart.

***

In the spring of 1987, Me-Tu delivered a son sired by Hugo who the keepers named Chang Dee (a bastardization of *chohk dee,* Thai for "good luck"). That was also the spring when the idea of enrichment arrived at the zoo.

Comprehensive animal care had evolved to account for not only the animals' physical comfort and health, but also their mental and emotional wellbeing. Any zoo seeking accreditation had to demonstrate a program geared toward providing variety and stimulation to its animals. In order to become accredited, a facility must be evaluated by recognized experts in the profession, who measure their findings against established standards. The Association of Zoos and Aquariums (AZA) has been the primary accrediting body for over forty years, although the Zoological Association of America (ZAA) also offers accreditation. Of the approximately 2,800 animal exhibitors licensed by the United States Department of Agriculture, fewer than 10 percent have AZA accreditation.

Enrichment is divided into six categories: environmental, habitat, sensory, food, social, and behavioral. Environmental enrichment involves natural or man-made objects that animals can manipulate safely. Habitat enrichment allows for a change of location and experience by utilizing the area's usable space. (This is not the same thing as total exhibit space, which may include areas inaccessible to the animal.) Sensory enrichment elicits species-specific responses to sound, touch, smell, taste, and visual stimuli. Food enrichment encourages hunting, foraging, and feeding behaviors by hiding meals within the habitat in order to motivate problem-solving strategies. Social enrichment creates opportunities for natural interaction within a species. In mixed species exhibits, it facilitates symbiotic or complementary activities between groups that would naturally occur in the wild. Behavioral enrichment provides mental stimulation and increases intellectual focus.

At first, the best Roger and his keepers could come up with by way of enrichment was to scatter various prunings around the yards, sticking them into holes in the ground or under chained-down logs so the elephants had to move around to find them as if they were browsing in the wild. Later came a shower system rigged with a pull-chain so the elephants could run it independently.

Despite these minor innovations, Roger admits that enrichment often fell by the wayside. He understood the logic behind it, and Haight was particularly excited by the possibilities, but the simple truth was that they were often so short-handed in the barn that it was impossible to accomplish anything except the daily basics of food, water, health care, and cleanliness. The elephant keepers frequently butted heads with those in the administrative and education departments, who couldn't seem to understand why they weren't able to do more.

One of the most innovative solutions to pachyderm lethargy and boredom was appleball, a game that involved a quantity of apples and a baseball pitching machine donated by the

Jugs Pitching Machine Company of Portland. The machine was set up behind a concrete retaining wall outside the elephant barn and adjusted to "pop fly" apples into the yard at random, rather than firing them in a line drive. This forced the elephants to give chase if they wanted to earn a treat.

Herd dynamics gave higher-ranking elephants the freedom to stake out those areas where most of the apples tended to land, which put the smaller, less assertive animals at a disadvantage. But Susi, the lowest of the low in the pecking order, learned to read the angle of the pitching machine to determine where the apples would land. She'd start running before the machine fired and arrive ahead of everyone else, outsmarting her competition.

In Roger's opinion, the most beneficial enrichment protocol was simple training. Regular and consistent hands-on instruction encouraged cooperation and built trust between the elephant and its handler. These weren't fancy circus maneuvers by any means—merely the basics he'd once used when performing with Packy: back up, kneel, lie down, hold still, trunk up, lift a foot, and accept leg chains.

It's worth noting that sometimes enrichment involves no activity at all. Zoo visitors are frequently disappointed by the lack of action in an exhibit and wrongly assume the animals are bored when in reality relaxing and snoozing are a normal part of their day.

In the summer of 1987, two young elephant keepers from Germany visited the zoo. Rudy and Felix were vacationing in the United States and wrote ahead to ask if they might tour the elephant barn. Roger was happy to accommodate them. He met them at the gate and escorted them across the zoo, struggling to overcome the language barrier. About one hundred yards short of the barn, both German keepers abruptly halted and raised their heads, scenting the air like a pair of bloodhounds. Rudy turned to Roger and, uncertain how to word his query, ran his index fingers down either side of his face from temple to chin, miming the oily excretions that occur during musth. Roger nodded and held up three fingers. Packy, Tunga, and Hugo were all in musth. The keepers' eyes bugged. Felix hesitantly suggested *not* going into the building might be a better idea. Roger chuckled and promised to bring them out alive.

Rudy looked dubious. "How you say, 'It makes my sphincter very tight?'"

Roger laughed. "Son, you can't say it any better than that."

Inside the barn, he brought Belle into the front room, introduced the young men, and easily coerced them into demonstrating their techniques for working with elephants. Later, he put each of the bulls through the ERD in a display of the device's versatility. The Germans were visibly impressed and returned home determined to have one built for themselves.

Roger occasionally utilized the ERD to temporarily confine musth bulls, but he advocates noninterference—coupled with gentleness and understanding—as much as possible.

"It's foolishness to hold them accountable for their actions because they're batshit crazy and can't do anything about it," he said during an interview in 2016. "No living organism would choose to be that gosh-darned miserable if they had any control over it." In fact, studies of the condition indicate that excessive domination of a bull may ultimately interfere with its instinct to breed, a finding of particular significance to any facility hoping to develop a breeding program. Male working elephants in Asia that have been severely and continually dominated will eventually refuse to mate.

The Oregon Zoo barn allowed each bull his own space, but when one or more came into musth simultaneously—a situation that occurred with what Roger described as "depressing regularity"—shifting the animals from one area to another became a thought-provoking, sweat-inducing game of musical chairs.

"The first would go into musth, wanting to break or kill whatever he could reach, and that would set off the next one, who'd set off the third," Roger said in 2016. "By the time he really got going, the first one, who'd come out of musth by that point, went back into it. We used to joke about who was craziest from the smell of testosterone, them or us."

With Hugo, he perfected a technique that allowed him to intimidate the rancorous bull without using the ERD. The interior walls of the barn didn't reach the ceiling, so whenever Hugo misbehaved, Roger would climb to the top of the wall and issue commands from on high. Something about the gravelly voice coming down from above convinced the bull Roger was much larger than him, making compliance preferable to battle.

Both wild and captive bulls have been observed to fast during musth. Roger's bulls would go through the worst of it "hardly eating enough to keep a sheep alive," he recalled. "You give it to them, but they just piss on it and waste it. Sometimes, they'd go ten days without eating. It was nerve-wracking because I could literally watch the surplus flesh slide off them." As soon as musth was over, though, their appetites returned, and they quickly regained the lost pounds.

One weekend in 1987, Roger and RoseMerrie left their girls at home and traveled over the mountains to Oregon's high desert country for a rare three-day visit with his brother and sister-in-law. They returned in high spirits with a truckload of firewood and good stories to share.

Michelle met them at the door and told her dad to call the zoo. Roger's three days of relaxation evaporated in a hot *whoosh*. He dialed the barn. When the receiver was lifted at the other end, he didn't even bother to ask what the problem was—he just told them he was on his way and hung up. Back into the truck he went.

Haight met him at the employee gate. As they walked across the zoo toward the barn, Haight explained that he and keeper Charlie Rutkowski had placed Packy and Hugo in adjacent rooms while they cleaned the barn and shifted animals. Hugo was in musth and ideally would have been situated two rooms from Packy so they wouldn't unduly antagonize each other, but since conditions are rarely optimal in the world of zoo keeping, they wound up in adjacent areas, separated by an elephant-sized hydraulic door beside which sat a smaller man-hatch keepers used when they didn't need to open the larger door. The bulls could hear and smell each other, but not see or reach each other. At least that's what the men thought.

Hugo battered the door with his head, trying to buckle it off its track, determined to get at his rival. When that didn't work, he and Packy took turns hurling macho insults at each other. The noise echoed through the barn, a background cacophony ignored by both the keepers and the thoroughly unimpressed cows.

Suddenly, an eerie squall unlike anything the men had ever heard cut the air. It lifted the hair on Haight's scalp and sent him and Rutkowski running. They found Hugo rocking side to side, sucking his bleeding trunk. Spying a section of severed flesh lying in Packy's manger, Haight reached past the bars, slid his fingers inside the nostrils, and lifted it out. Rutkowski puked.

They alerted Dr. Schmidt and moved Hugo into the ERD where he could be examined. There he'd remained, receiving antibiotics and painkillers every twelve hours.

Hugo was notoriously brilliant, a veritable pachyderm Professor Moriarty. The keepers theorized that when his attempts to get at Packy proved unsuccessful, Hugo inspected every inch of the wall and discovered the man-hatch. Using his dexterous trunk like a hand, he worked the door open. While most elephants' trunks are about six feet in length, Hugo's was closer to seven. Those additional twelve inches allowed him to snake through the opening into Packy's enclosure and reach the manger. He was busily inhaling olfactory clues when Packy spied the intruder, rushed over, and bit down, crushing the flesh and severing a section longer than Haight's hand.

Inside the barn, some of the staff stood beside the ERD. Ignoring their greetings, Roger pushed past to get a better look at his bad boy. Securely and comfortably confined by the barred walls (if not altogether happy about it), Hugo glared straight into his eyes, defiant as always.

Meeting the elephant's gaze, Roger spoke over his shoulder to those assembled. "He's hurting, but don't any of you get within reach because this old bastard would like nothing better than to drag you in there."

Hugo remained confined for over a month. Roger acknowledged this was a less than optimal solution, but it was the only one that allowed the bull to receive the care he required without endangering his handlers. Having been exposed to the ERD on a daily basis since his arrival in Portland, Hugo tolerated confinement better than one might imagine. His injury made it impossible for him to feed himself, so he relied on keepers for assistance. Every day, he received a bale of hay, one handful at a time. For grain, an ingenious contraption was constructed out of four-inch PVC pipe outfitted with a plunger. When the filled tube was presented to him, Hugo opened his mouth to receive it and the plunger was depressed halfway. Once he'd chewed and swallowed that portion, he opened his mouth again and the second half was dispensed. "You could tell he was grateful," Roger remembered.

Hugo also learned the difference between cold and warm

water. When cold was delivered via a hose, he knew he could drink it. When warm salty water was brought in a container, he learned to suck it in, hold it as the elephants did for their periodic tuberculosis testing, and then blow it out, flushing the wound.

Six weeks into the bull's convalescence, Roger and Haight paused on their rounds one day to offer him hay. Hugo accepted it, spat it onto the floor, and picked it up with his truncated proboscis. Placing the hay in his mouth, he chewed slowly, eyes on the men.

They were thunderstruck. None of the keepers had known whether he would ever fully recover, but Packy's bite had severed Hugo's trunk at such an angle as to leave it longer on top than on the bottom, creating a little nub of flesh. That tip wasn't nearly as dexterous as the one Mother Nature normally bestowed, but it was enough for Hugo to work with, and he learned to use it well.

On the day he was released from the ERD and allowed outside, Hugo emerged from the barn and went straight to the wall, rubbing his body against it to loosen six weeks' worth of dry, itchy skin. Haight brought out a long-handled, stiff-bristled broom, asked him to lie down in front of the opening between the yard and the barn, and rubbed him down with it. Hugo actually groaned with pleasure.

Sometime later that day, Hugo sank into the pool for a long, luxurious soak. Wallowing in bliss, he ducked his head and sprayed a trunkful of water across his back, the pink mottling of his ears bright in the sunlight. It remains one of the most perfect moments Roger has ever witnessed.

Chapter *Eleven*

HARD TIMES

1988–1991

On June 18, 1988, Roger walked his firstborn daughter down the aisle at her wedding. Memories flashed through his mind: the red-faced newborn who scared him so; the little girl bossing her younger sister around; the six-year-old hunter with muddy boots tracking along behind him through the forest; the teenager with a runny nose sharing coffee and sandwiches with her old man in the elephant barn; her first date; her first heartbreak; the struggles she'd endured; and the triumphs that defined her life. He was immensely proud of both his girls and loved them with all his heart.

The emotional high from the wedding lasted several days as out-of-town visitors stayed to catch up with one another. Roger cashed in a few precious vacation days to enjoy the reunion, but the whole thing was ruined on the last day when he received a phone call telling him Susi had died.

The star player of appleball had been at the zoo for fourteen years. Despite numerous attempts to breed her with Packy, Tunga, and Hugo, she failed to conceive. Someone at the zoo had decided it was time to try a fertility drug, to which Susi had a fatal reaction. Like Thonglaw in 1974, she succumbed within moments.

Her death haunted Roger. The shy, inoffensive cow had never bothered anyone, never caused a lick of trouble in the barn, and spent her days ghosting along the edge of the herd, unwilling to impose or draw attention to herself. He felt guilty for not being there when she died, for not being available to at least offer a token voice of dissent before she was injected. He believed he'd let her down, just as he had Tuy Hoa—nothing could convince him otherwise.

With Susi's death and the departure of Hugo's son, Chang Dee, to Ringling Bros., the population settled at eleven: Rosy, Belle, Pet, Packy, Me-Tu, Hanako, Tamba, Tunga, Rama, Sung-Surin, and Hugo. Then, several months later, Cindy arrived.

Captured as an infant at least thirty years prior, Cindy spent her early life in traveling sideshows performing at shopping centers. Point Defiance Zoo acquired her when she was three years old, and she spent the next seventeen years without another elephant to teach her and help her grow into a calm, emotionally balanced adult.

Roger first met her when zoo curator Roland Smith—the same Roland Smith who'd driven with Roger to Canada to deliver Tina to the airport—asked him to evaluate Cindy because just looking at her made his mouth go dry. Knowing Smith was not easily daunted, Roger drove north to take a look. He stood outside Cindy's enclosure not saying a word, asking nothing of her, merely watching. She glared at him and thrust her trunk past the bars of her enclosure, making it clear she wanted nothing more than to knock him down. Afterward, Roger told Smith that his dry mouth was the Almighty trying to keep his silly ass alive. "I've never met an elephant that gave me such a bad case of chills," he said.

As her aggression increased, Point Defiance hired an animal trainer to work with her. When that didn't produce the desired results, San Diego Wild Animal Park took custody of her. Cindy arrived in California in 1982 and remained there for seven years. She refused to mate, and worked grudgingly with the keepers and other elephants, her behavior anything but cooperative. One trainer thought her intelligent, but bored. Her rampages continued and at least two people were injured. *The Los Angeles Times* reported witnesses coming to her defense, saying she was striking back after a keeper beat her.

Point Defiance was willing to take back the troublesome cow, except they were in the midst of building a new $2.3 million elephant facility complete with improved yards, a pool, larger rooms, and a hydraulic ERD that would make handling Cindy safer for all concerned. When Roger learned the Oregon Zoo offered to house her in the meantime, he threw his hat against the wall. This would be like Hugo all over again: a brilliant animal with a personality irreparably damaged thanks to the treatment she'd received at the hands of some thoughtless humans. He couldn't blame her for thinking the entire race was evil, but any kindness she received now was a case of too little, too late. When she arrived in November 1989, the first thing Roger did was warn his crew to not take chances. "She'll kill you so quick, there won't be time for 'oh, shit.'"

A few months later, a routine examination of Belle's right elbow revealed that her refusal to use the limb in a normal manner had caused adhesions to develop, rendering the joint completely immobile. Roger wasn't surprised. Belle refused to lie down as healthy elephants do from time to time, and slept leaning against a wall or fence. She held the frozen limb away from her body and swung it counterclockwise when walking. The change in gait altered the balance of weight on her other feet, causing abnormal wear to her pads and nails. Treatment continued as before: anti-inflammatories, foot soaks, antibiotics when necessary, and frequent checks to keep her feet in the best shape possible. Roger tried not to think about Tuy Hoa and how much this situation reminded him of her.

The elephant barn's bad luck continued. In September of 1990, Me-Tu shoved Tamba into the dry moat. The eight-foot-deep moat was designed in such a way that any elephant who slipped into it could walk to a gate and be let back inside the barn. Unfortunately, Tamba had developed an arthritic leg, preventing her from getting up on her own. Not only that, she was lying crosswise, scrunched against the wall on either end with little room to maneuver.

Roger stood looking down at her, sighed, and lowered himself into the moat. For the next three hours, he kept her company until a rescue team arrived with a pair of tow trucks, a sling, and some inflatable airbags. They attached these to her midsection, then to truck cables. In a beautifully orchestrated move, air was slowly pumped into the cushions as the sling cables lifted. In a relatively short time, Tamba was back on her feet.

Throughout the operation she remained quiet and cooperative. Having suffered nothing but minor scrapes and bruises, she became quite talkative afterward, as if wanting to share the details of her adventure with Roger, her favorite human, as they walked together to the moat gate and were let back into the barn. Roger recommended she be moved into the other herd, away from Me-Tu, but was told to leave her where she was.

Having had his concerns brushed aside so often in the past, he should have been used to it, but time didn't make the dismissals any easier to swallow. He never thought his way was the only way—far from it—but his opinion ought to at least carry some weight.

"I wasn't looking to be right, and didn't give a damn about who got credit," he said in 2015. "What I cared about was what works, what's safe. It sticks in your craw when you're getting told by your so-called superior that he wants this or that done and you know damn well it's the wrong way to go, but there's nothing you can do about it."

Female elephants in the wild typically live out their lives as part of the matriarchal group to which they're born. For males, however, something different occurs. As young bulls begin puberty, the emergence of challenging behavior results in their being forcibly evicted from the herd to assume the largely solitary life of an adult. Rosy's young son, Rama, reached that stage in 1990 and his snotty attitude was met with shoves, kicks, and trunk thumps as the cows attempted to drive him out of their company. The problem was, they were locked together in an exhibit and the juvenile bull had nowhere to go.

Roger recommended Rama be transferred into an area of his own, though privately he wondered how the keepers would juggle four adult bulls when the three they had were already straining everyone's creative ability. The curator and veterinarian denied his request, believing the seven-year-old bull still needed time with the herd and would benefit from their influence. A few weeks later, the cows ganged up on Rama and nearly shoved him into the dry moat.

This time, Roger was more forceful in his suggestion. He met with the curator and veterinarian face-to-face so there could be no misconstruing his intent. He began calmly enough, outlining his reasons in a rational manner. When that didn't gain their support, he explained again, but with rising emotion. Still no success.

Roger let his feelings fly. Snatching the campaign hat from his head, he flung it to the floor in frustration and yelled that nothing good would come of the bull being left in with the herd. The tantrum gained him nothing; the answer remained no. He stalked back to the barn in a black fury.

Shortly after that meeting, his prediction came true. Having irritated the cows past tolerance, Rama was shoved into the dry moat and suffered a greenstick fracture in which one side of the bone breaks and the other side bends. The serious injury didn't heal as well as was hoped, and Rama was left with a permanently altered gait that redistributed his weight onto three legs instead of four. Roger didn't have to say "I told you so." They all knew it.

Those in charge finally stepped up to make certain a fall like this would never happen again, and the outdated moat was filled in at long last. Roger was grateful, but that didn't mean he forgave anyone for the ham-fisted way Tamba and Rama had been handled. These managerial conflicts might have tempted another man to resign, but that wasn't Roger's style. His commitment to the elephants was dogged, determined, and ceaseless. Rather than drive him away, confrontation only hardened his resolve to remain their advocate.

In October 1990, two new employees arrived in the barn. Sioux Strong and her husband, Fred Marion, were animal keepers and trainers from California's Santa Barbara Zoo where they'd cared for Asian elephants Mac and Suzi. Strong possessed a competent, easy manner that immediately impressed her new boss. Marion was energetic and worked hard. Roger liked them both right from the start and they quickly became two of his favorite people.

"Freddy was a dynamo, easily pound for pound the toughest little bugger I ever encountered, and he'd do anything in the world for you," Roger recalled. "It didn't matter what the work was, Fred was the guy to go to if you needed help. We clicked right away, same sense of humor, same work ethic. Now, Sioux. If something had a heartbeat and respiration, Sioux could train it. I never met anyone who could match her ability. She's like a third daughter to me and I love her dearly."

The research Dr. Rasmussen hoped would isolate the sex pheromone associated with estrus was still going strong in 1990.

In addition to collecting and analyzing urine and blood (and, in a related study, samples of the temporal secretions of musth bulls), the elephants participated in a daily "sniff test." The test utilized adjacent rooms with a connecting door. A bull went into one room, and a cow into the other. The cow was backed up to the door, and the door opened just wide enough to allow the bull's trunk access to her hindquarters. He'd check her urine and urogenital secretions, and an observer would note the bull's response—length of interest, lack thereof, or penile erection. Each test took seconds to complete.

Once the cows understood what was going on, they cooperated without issue. The lone exception was Tamba. For reasons Roger never understood, she truly hated the entire experience. Perhaps it left her feeling vulnerable, being unable to avoid the bull behind her. She never offered her handlers any resistance, but anyone with an ounce of empathy could tell from her sad eyes and woebegone expression that she was absolutely miserable the entire time.

On December 6, 1990, two short months after the arrival of the Santa Barbara keepers, Roger and Marion were leading Tamba in for her dreaded daily inspection when they met Hanako, Haight, and another handler coming the other way. The threshold was certainly wide enough to accommodate both elephants with room to spare, but Tamba suddenly wheeled. She collided with Marion, slamming him sideways and crushing him between her and the cement wall.

Shouting, struggling, and squealing are all Roger recalls of the frantic moments that followed as havoc echoed through the barn. An ambulance rushed Marion to the hospital as the keepers returned Hanako to the herd. Tamba was put in temporary isolation pending evaluation because no one understood what set her off. Despite later comparing notes with the others, Roger never did know with any certainty. At the time, a local newspaper reported that prior altercations with Hanako made Tamba fearful of the larger, more dominant cow. She'd borne the brunt of abuse from Hanako and Me-Tu for years. A desire to escape close proximity to her tormentor may have caused Tamba to panic and accidentally run into Marion, but at least two individuals who were present that day believe she disliked Marion for reasons unknown and purposely attacked him.

Leaving the barn in the capable hands of his crew, Roger sped to the hospital where he was joined by RoseMerrie. Strong and her mother were en route to California by train, having just arrived at their destination when word of the accident reached them. Oregon Zoo Director Y. Sherry Sheng, who had replaced Warren Iliff's successor Gene Leo in the '80s, wired the funds so they could immediately fly home.

Roger frantically paced outside the emergency room for what felt like hours as patients, doctors, and nurses came and went. He stopped every person wearing scrubs to inquire about Marion's condition, but no one seemed able or willing to tell him anything. When Strong and her mother arrived, and she introduced herself as the injured man's wife, she had no better luck. It was only when Roger threatened to kick open every door until he found Marion that a senior doctor allowed them in to see the injured man for a couple of minutes.

Marion lay on a gurney, unconscious, the back of his head split open and blood everywhere. Seeing his voluble friend unresponsive and vulnerable leached the fire from Roger's soul. Laying a hand on Marion's wrist, he gave it a gentle squeeze to let Freddy know he was there and asked the doctor whether he would live or die. The answer was noncommittal; they had to stabilize him and run tests before they would know.

Marion proved to be tougher than anyone realized. He survived the incident, made a full recovery, and returned to work with the elephants again, including Tamba.

Strong supported her husband through that dark time and his lengthy rehabilitation, but she lived for Fridays, when she and Roger teamed up in the barn. Estranged from her father, she was drawn to her gruff boss, his down-to-earth attitude, and his loving dedication to the elephants. Their banter was gently abusive but friendly, their conversations often deep and thought-provoking, and their silences companionable. It was the perfect working relationship.

They were chatting amiably one day in January 1991, moving from task to task, when a mysterious *skritch, skritch, skritch* caught their attention. When they went to investigate, they discovered Hugo wandering around his breakfast-strewn enclosure with the handle of a paring knife in his trunk, methodically scraping the blade against the concrete floor. How the knife came to be in the cage they didn't know, but it was safe to assume it arrived in one of the fifty-five-gallon drums of produce regularly donated by several area grocery stores. Of greater importance was how to get it away from Hugo. They couldn't walk into the cage and just take it as they would with one of the cows—to do so would be suicide.

Strong studied the bull for a few moments, watching closely as he meandered from one place to another. She nodded decisively and announced that she could get him to surrender the knife.

Roger snorted in disbelief and stalked away, calling over his shoulder to come find him when she gave up.

He was in the office catching up on paperwork when Strong walked in and dropped the knife on the desk in front of him. He stared at it and glanced at his wristwatch. Eight minutes had elapsed. "What did you do to get it?" he asked. "Kill him?"

No, Hugo was very much alive. What Strong had done, in true Tucker fashion, was offer the elephant a better deal. Having observed he periodically dropped his newfound plaything in order to investigate a piece of food before picking it up again, she'd taken some quartered apples from the feed room and stood in front of his cage waiting for him to drop the knife as he browsed. When he reached to pick it up again, she crooned an approving "*Good*," and tossed him a piece of apple. If Hugo showed an inclination to ignore the knife, she tossed the enticing tidbit closer to it than to him and praised him when he approached.

It didn't take Hugo long to figure out the rules of this game: something about this new toy won him a delicious treat. When his behavior with the knife demonstrated that he understood, Strong upped the ante. She turned her back, pretending to ignore him as he picked up the knife. This was a twist to the game Hugo hadn't anticipated. If she wasn't watching him, she wouldn't know when to toss a chunk of apple.

Within moments of turning her back, Strong heard a *clunk* behind her and turned around. Hugo had tossed the knife out of the cage and was waiting with an expectant expression. She praised him effusively and rewarded him with the remainder of the apples, then picked up the knife and went to find her disbelieving boss.

When she finished her tale, Roger sighed with resignation. "What's my lack of faith in your abilities going to cost me?"

Strong smiled. Somewhere in the world, there exists a picture of Roger, flat on his back in the west yard, making a snow angel.

***

Roger's relationship with his dad improved over the years as Leonard's personality mellowed with age. The elder Henneous was a gentle and pleasant father-in-law to RoseMerrie and delighted in his role as grandfather, teasing the girls and buying them candy. He rarely visited Roger at work because the zoo's hilly terrain made walking difficult, but on those occasions when he did, he stood by grinning and shaking his head in amazement at his son's ability to work with the elephants.

If anything bothered Leonard, it was the way his aging body began to betray him, refusing to let him do the things he used to do. His mother died when he was nine years old and he'd labored hard every day since. To no longer be able to climb ladders, crawl under a car to change the oil, or even rake the yard felt like the ultimate treachery. It hurt his pride to ask his sons for help, despite their willingness.

For years, Leonard had been dogged by a persistent cough, but he always brushed it off and refused to see a doctor. It wasn't until July 1991, when he began to experience shortness of breath, that he made an appointment and learned he had emphysema. Besides being a lifelong smoker, his lungs had been further compromised by grain dust on the farms he'd worked and rubber dust at the Portland tire company where he'd been employed for over twenty years.

The news shocked the family, as did the doctor's prognosis: Leonard likely didn't have long to live. In an effort to better assist his parents—particularly his mother, who was having difficulty accepting that the man she'd slept beside for over fifty years would soon be gone—Roger stole minutes from the zoo and home, his back always covered by coworkers in one area and RoseMerrie and their daughters in the other.

Leonard declined rapidly and it broke Roger's heart to see him turn into an old man seemingly overnight. The family made the painful decision to move Leonard into a nursing home where he could receive round-the-clock skilled care. No one was happy about the situation and the job of admitting him fell to RoseMerrie. The mere idea of placing his dad into care almost made Roger physically ill, and he despised himself for not being the one to do it. As the eldest, he felt it was his job.

Later that month, Tunga was sold to the Cuneo Corporation. Although the Oregon Zoo had hoped to breed him, he'd displayed neither interest nor aptitude, siring just one calf, Look Chai, in twelve years.

On October 6, 1991, three months after Leonard's terminal diagnosis, Pet went into labor with her sixth calf. By now an old hand at this baby-making business, she'd breezed through the past twenty-two months with nary a flutter, entirely unperturbed by the life growing inside her. As she entered advanced labor with Belle rocking and crooning beside her, excited zoo personnel gathered to witness the event. Roger stood close to the bars, grateful something good was about to ease the strain of Leonard's illness.

Pet grunted and squealed with effort as Belle ran her trunk across her old friend's body, offering comfort and support. When at last the baby slipped free and hit the floor, those gathered to celebrate stared in silent dismay. The calf's skull was misshapen; its eyes appeared too flat, and one leg was deformed. To their amazement, it stirred and tried to stand, but couldn't. Rather than assist with their feet and trunks as usual, Pet and Belle simply turned and walked away.

"It was the saddest thing I'd ever seen," Roger remembered in 2015. "They never displayed any interest at all." As the cows began to feast on a fresh pile of hay, keepers entered the cage to retrieve the calf. They discovered an opening in its spine indicative of spina bifida, a condition in which a portion of the spinal cord is left exposed due to the incomplete closure of the backbone and the membranes surrounding it.

Keepers lavished love on the calf for the few hours it survived, but the veterinarians determined there was nothing they could do except let it go. Pet and Belle, with their greater wisdom, had understood the unavoidable outcome from the beginning and paid no attention as they munched their hay.

Chapter *Twelve*

ELEPHANT WOES

1992–1996

Not long into 1992, Roger's request to move Tamba to the other herd came back to haunt him when Me-Tu challenged her once more. This time the altercation occurred in the front exhibit room. By the time keepers were able to break it up, Tamba had sustained a greenstick fracture in her right front leg. Dr. Schmidt did what he could for her, and everyone hoped the injury would heal given sufficient time.

Roger, Haight, and the rest had their hands full with eleven elephants, so no one was very sorry when the Oregon Zoo received word from Point Defiance Zoo that their new elephant barn was ready and they were prepared to resume custody of Cindy. She was loaded into a truck without fanfare and wished the best of good fortune in her life in Washington.

The loss of Pet's spina bifida calf lingered with Roger, haunting his waking hours and his dreams. Calves had died before and he'd mourned each loss, but this one seemed particularly wrong, perhaps because it involved two of his favorite elephants: his pigeon-toed, tobacco-chewing partner and the world's best auntie. He obsessively replayed the event in his mind, unable to shake the image of Pet and Belle walking away from the doomed calf as though it didn't even exist.

As was his habit when bad times fell, Roger locked down his emotions and moved through his days with leaden determination. Most mornings, he was the first in the barn. After a full day's strenuous labor, he'd stop by his parents' place to check on Myra, then head to the nursing home for a visit with Leonard before returning home to RoseMerrie. Sometimes Melissa was there, sharing news of her life. Michelle stopped by often, waddling through the door heavy with her first pregnancy, her September due date quickly approaching. After dinner, Roger would pretend to read or watch television, or sometimes go back to his parents' house to mow their lawn or handle any odds jobs Myra couldn't attend to on her own.

On April 11, 1992, nine months after his diagnosis and six months after the death of Pet's calf, Leonard Elwin Henneous died at the age of seventy-six. The family was devastated, but no one more so than Roger. He and his dad clashed in the past, and in many ways didn't understand each other, but this was *Dad*—staunch, stalwart, hardworking, unflinching, demanding, dependable, frustrating, but always there. And impervious, or so the child still alive deep inside Roger believed. How could Dad possibly be dead?

The day of the funeral, Roger moved through the proceedings in a numb haze, going where he was directed, doing as he was told while Myra was ably cared for by her daughter and daughters-in-law. At the reception afterward, one of Leonard's old friends approached Roger. "Goddamn," he said as they shook hands. "You don't look eight feet tall to me."

"Beg pardon?"

"Your dad. To hear him talk, you could walk on water."

Roger stood pinned to the spot. He'd known Leonard was

impressed by his work with the elephants but never realized he'd been spoken about, let alone with pride. He thought his heart, creaking under the weight of grief, would break clean open. If only those words had come directly from his dad and not through some stranger.

As summer 1992 approached, the air grew gentle with the scent of mowed grass and turned earth. Flowers emerged, buds swelled on the trees, and the city's famous rose bushes put out their first leaves. The days grew heavy with heat and crowds of visitors filled the zoo. On days off, Roger busied himself in the backyard, planting tomatoes and RoseMerrie's beloved cucumbers and garlic. They went on vacation, but if pressed for details, he couldn't remember much about it. Already prone to depression, and now consumed by sorrow over the deaths of his father and the spina bifida calf, Roger sank into an emotional abyss.

In the first week of September, the family mustered its emotional resources to celebrate Myra's seventy-ninth birthday. Leonard's absence was keenly felt, but everyone tried their best to make the occasion a cheerful one. A week later, RoseMerrie stopped by to visit her mother-in-law and Myra mentioned offhandedly she hadn't been feeling well. She could offer no specifics, only that something didn't seem quite right. Hypersensitive to such things since her bout with cancer, RoseMerrie wasted no time in scheduling a doctor's appointment. The diagnosis was grim: breast cancer so advanced there was nothing Myra could do except go home and put her affairs in order. When Roger heard the news, he went off on his own and wept where no one could see.

The only bright spot on his emotional horizon was the arrival of another baby, this one of the two-legged variety—their granddaughter Victoria. Roger embraced the role of grandpa and readily babysat even if the baby was sick. Much to Michelle's amusement, he took careful notes on milk consumption, naptimes, and bowel movements just as he would in the barn. Victoria made him as close to happy as he could get, but not even her drooly grin could lift the cloud entirely.

Christmas that year was a tender and painful affair. Myra's cancer advanced so rapidly that, like Leonard, she was placed into adult care—much to Roger's heartbreak. As the year turned, he and RoseMerrie quietly marked their twenty-eighth wedding anniversary, hoping 1993 would keep its nasty surprises to a minimum.

On the morning of January 28, Roger walked into the barn and found Rosy down on the floor, unable to rise. All the foot soaks, antibiotics, and anti-inflammatories in the world made little difference, and her foot problems had caught up with her at last. Still, her keepers refused to give up. They removed the other cows from the area to create space to work and made several attempts to get her on her feet. Rosy just lay there, rolled onto her chest, and refused to move. Even Sioux Strong, with her renowned ability to get an elephant to do anything but fly, had no success. Rosy watched them with patient, wise, loving eyes, and told them what they already knew. It was time to say goodbye.

Roger balked at the notion. This was *Rosy*, for Christ's sake, Portland's first pachyderm, the gentle, sweet animal one of the keepers described as "God's perfect elephant." The barn without her was unthinkable. But he'd witnessed horses and cows down like this; hell, he'd seen antelope and bison, too, and God knows how many others in twenty-five years at the zoo. He knew when further effort was useless, but still he tried. Once last more, he begged her to rise. Rosy blinked and lifted her trunk to snuffle his chest, but did not move. Roger sighed.

Strong told him to wait a goddamn minute before calling the vet and rushed next door to the AfriCafe. Storming through the front door, she strode into the kitchen area, her boots practically striking sparks off the floor, and brazenly picked up a serving tray of day-old pastries. In a voice meant to stop any argument, she said, "I'm taking these to the elephant barn." In a final act of love, tears streaming down her face, she sat beside Rosy and fed her one pastry after another until they were gone.

Many hours later, when the deed was done and the horrid task of disposal completed, Roger drove home, opened a fresh bottle of rye, sat down at the kitchen table, and "cried, cussed, guzzled, and puked." He'd worn protective gear during the foul process and there wasn't a spot of blood on him that he could see, but the smell clung to him like a curse and condemnation. He wondered if Rosy's death had finally pushed him over the edge he'd balanced on for so long.

Eventually, he staggered into the bathroom to wash away the day, knowing he could never escape the memories. As he pulled his T-shirt over his head, a blood clot the size of his hand peeled loose from the back of his neck and landed on the rug. He stared at it and laughed, the sound flat with pain and gallows humor. The smell of blood hadn't been his imagination. Maybe he wasn't nuts after all.

Without needing to be told, his close-knit his family of keepers understood something was very wrong with Roger. Closing ranks around him, they bulled through their days, taking on his chores when they could. Herd dynamics shifted over the following months as several of the cows contended for the position of matriarch in Rosy's herd. It came as no great surprise when Me-Tu ascended her mother's throne.

On March 17, 1993 six weeks after Rosy's death, Myra Pearl Stanton Henneous died. Roger handled his grief as he always had. He buried it deep inside, letting it out only when no one was around to witness his tears, unwilling to burden his family with his emotions when they were dealing with their own sorrow. He knew his closest friends and coworkers would have willingly listened and offered support, but he didn't know how to express his anguish. It wasn't only Myra—it was also the loss of Rosy and Leonard, Susi and Tuy Hoa. He buried his woe beneath work, where it festered. Labor became a panacea to bandage his soul, but couldn't protect it from further hurt.

The greenstick fracture Tamba sustained a year earlier never properly healed and she showed little sign of improvement. Roger's heart ached to see her limping from one place to another. Medication somewhat eased her pain, but now she began to use her trunk like a crutch, inching her way around her enclosure. There were no other means to alleviate her discomfort, and only one way to give her peace.

Five weeks after his mother's death, Roger wept as Tamba was euthanized. Titanic guilt engulfed him. Here was another death laid at his feet, another miracle he'd failed to manifest. His soul railed at the Universe, demanding it explain what his purpose was here on Earth if not to protect and care for his family and his elephants.

Those who loved Roger understood he was depressed, and maybe even guessed it went deeper than that, but he was adept at keeping the weight of his rage and despair a secret. On his darkest days, he stood outside Hugo's yard, watching the bull pace or wrap his trunk around an immense tree stump, swing it high overhead, and slam it against the ground before grinding it into the soil with his forehead. Suicide by elephant seemed a good way to go. The deed would be over in a flash, considered a tragic job-related accident, and the insurance money would guarantee RoseMerrie and the girls were well taken care of for the rest of their lives. So deep was his despondency, it never occurred to Roger anyone might actually mourn *him*.

In the end, he couldn't go through with it. If he let Hugo kill him, the bull would pay the ultimate price. Looking back on that dark time, Roger said, "I couldn't justify letting him take the rap and be destroyed because of my own cowardice."

Summer 1993 arrived with its usual bustling crowds of visitors, autumn brought rain and field trips with school children, and winter slid in on a patina of ice. On New Year's Eve, Roger and RoseMerrie celebrated their twenty-ninth wedding anniversary, and Roger began to hope the run of bad luck was over.

That cautious optimism lasted until March 1994 when routine foot maintenance revealed swollen and inflamed areas between the nails of Belle's left foot several weeks after the surgical removal of damaged tissue caused by the impingement of adjacent nails. Radiographs showed no bone abnormalities, and the areas were treated with antibiotics and anti-inflammatories.

For the next several months, Roger divided the bulk of his attention between Belle and Me-Tu, who was pregnant again. Like her mother Rosy before her, Me-Tu mated readily, carried without incident, delivered healthy calves, and cared for her children in the best elephant fashion. No one expected things to change, but on the afternoon of August 30, 1994, she went into labor two months shy of full-term.

No one was particularly concerned about the early delivery. Advances had been made in the field of elephant obstetrics and Dr. Rasmussen's team finally achieved success in isolating the pheromone they'd so diligently pursued, but so much about elephant gestation remained inexact. Perhaps they'd judged the date of conception wrong. Gestation was typically 640–660 days, but there were cases of elephants producing healthy calves on either side of that window.

Staff convened to watch as usual, eager to greet the newcomer. They remained in the barn all night, taking turns sleeping on hay bales stacked in the keeper alley. At around eight-thirty the following morning, Me-Tu grunted, pushed, and out popped what Roger described as "a little peanut of a calf." He looked from Me-Tu's belly, to the feisty little female already struggling to stand, and back again. "No bigger than she is, there's room in there for one more," he said.

Dr. Schmidt shook his head. The incidence of twin births in elephants was so rare as to not warrant consideration, occurring in less than 1 percent of pregnancies. In his opinion, there was nothing left inside Me-Tu except placenta.

Roger remained unconvinced, not because he thought he knew more than the vet (although the idea had occurred to him once or twice), but because Me-Tu was telling him something wasn't right. As the day progressed, she swayed from side to side as though trying to shift an internal load. If there was another calf in there, it must be a transverse presentation—the equivalent to a human baby being born feet-first—or laying crosswise in the uterus, unable to budge.

He called Schmidt's office and suggested a pelvic exam be performed, but the vet repeated his comment about the placenta and refused to argue the point. Irritated at Schmidt's dismissal, Roger slammed down the telephone and returned to his spot outside the cage, watching Me-Tu sway as midwives Belle and Hanako divided their attention between the tiny newborn calf and her fretful mother.

Sioux Strong paused during her rounds to see how things were progressing. As she and Roger watched, Me-Tu suddenly rocked back on her hind legs and placed her front feet high against the wall. The posture was one they'd never seen an elephant in labor assume. Me-Tu vocalized, clearly in pain, as Belle and Hanako crooned in sympathy and stroked her with their trunks.

Something was obviously wrong. Roger telephoned Schmidt again to report the strange behavior and received the same response as before: there was nothing inside Me-Tu but placenta.

Twelve hours later, after staff had gone home leaving Roger in the barn with night keeper Tim Brooks, the second calf arrived. In a chilling replay of Pet's performance three years earlier, Me-Tu and her midwives simply walked away the instant it hit the ground. Seated on a hay bale, Roger stared at the inert form. It had been a rare set of twins after all. He felt no sense of vindication at being right, only heartbreak and loss. What a wonder this would have been had they both survived.

The calf blinked.

Roger shot to his feet, screaming for Brooks, and they dragged the unresponsive infant into the hay room. While the night keeper began CPR, Roger ran to the office, snatched up the telephone, and dialed Schmidt's home number. "You know that placenta you were so sure about?" he roared when the receiver lifted at the other end. "Well, it's got a trunk and two eyes and if you move your goddamn ass you might even find her alive when you get here!" He slammed the telephone down as hard as he could and went to assist Brooks.

Schmidt and his wife Anne Moody broke the speed limit getting to the barn, but it made little difference. The calf died two hours later. A necropsy revealed brain damage and a collapsed lung due to protracted labor. Schmidt's unwillingness to consider the possibility of twins was a tragic error on his part and one Roger never forgave.

Even if they'd understood what was happening, it's uncertain whether they would have been able to shift the calf inside its mother enough to change its position. Surgical options such as C-sections are considered lethal in elephants due to the dimension and anatomy of the animal's reproductive tract. A two-foot incision secured by sutures cannot support several thousand pounds of weight. The few elephants that have undergone a C-section all died shortly thereafter due to complications from the surgery.

Keepers christened the surviving twin Rose-Tu, after her grandmother. From the moment her feet hit the floor, she was a spirited creature, displaying aspects of her strong-willed matriarch mother, Me-Tu, and battle-prone father, Hugo. She entertained spectators and staff with her antics, racing throughout the exhibit room with her ears and trunk flapping, tail stuck straight out like a bottle brush. She even made Roger laugh, though he couldn't look at her without also seeing her dead sister. He needed all the laughter he could get.

Sioux Strong departed the zoo in October 1994, her marriage to Fred Marion having dissolved. Roger felt torn between his friends and hated like hell to see her go, but he wished her well and knew they would remain in touch.

By 1995, every elephant in the barn except for little Rose-Tu suffered from one sort of foot ailment or another. The keepers dispensed old treatments and telephoned professionals in the field asking for advice on new approaches. Every day, Roger felt the weight of ghosts on his back, reproaching him for what he could not seem to accomplish no matter how hard he tried.

An ulcerative lesion between the toes on Belle's left foot began to spread despite keepers treating it with a variety of topical medications. Corrective trimming revealed a larger area of infection than expected. Belle was sedated and Roger stood beside her head, one hand tucked into the soft area behind her ear as veterinarians performed a deeper debridement, followed by treatment with more antibiotics and analgesics. She wore specially designed leather sandals on her front feet in an attempt to keep the wound as clean as possible. For the better part of a year, Roger and his crew held their own against the encroachment of infection.

It all went to shit in February 1996. Eighteen months after the birth of her twins, Me-Tu succumbed to the ravages of foot rot. Watching the light in her eyes fade, Roger felt numb, but at the same time overcome with agony, as if his soul were full of broken glass. In slightly more than four years, he'd lost Leonard, Myra, the twin, Rosy, Tamba, and Me-Tu.

The hardest blow, of course, fell to little Rose-Tu. Elephants communicate constantly among themselves using infrasound.

From the moment Rose-Tu was born, her mother's voice foremost in her mind, filling the calf's entire body with its familiar hum and pulse. Now there was silence.

To no one's surprise, Belle stepped in to fill the void and became the stabilizing adult influence Rose-Tu so desperately required. With the able assistance of fourteen-year-old Sung Surin, Belle comforted the calf and shared her grief. More importantly, she nurtured Rose-Tu, defended her, and taught her what it meant to be an elephant.

In April 1996, a follow-up radiograph of Belle's foot again showed no bone abnormalities. Topical and intermittent systemic antibiotics were prescribed and continued until November 1996, when a third radiograph revealed bone disintegration. Roger's brain reeled, wondering what, if anything, they could do to stop it. Veterinarians prescribed antibacterial foot soaks, topical medications, anesthetics, and antibiotics.

One month later—despite all their efforts—the infection in Belle's foot spread further, encompassing more than one bone. Debridement under sedation became more aggressive, with care taken not to undermine the foot structure and make her erratic stride worse than it already was. Bacteria remained resistant to treatment and the limb's circumference was measured at several places to monitor swelling. For the first time since Belle's foot issues were discovered, the zoo released information about her condition to print and broadcast news agencies. Zoo medicine and surgical specialists at the UC Davis School of Veterinary Medicine in California were brought on as consultants and advised amputation of the infected bone.

When he heard the news, Roger wandered out into the barn and slipped between the bars of Belle's cage to tell her the news. She accepted it without emotion, which was just as well, because Roger was a wreck. "Surgical amputation," he muttered into her ear. "How the hell are we supposed to accomplish that?"

Chapter *Thirteen*

BELLE

1978–1979

Every elephant at the Oregon Zoo had its share of fans. Packy, Rama, and Hugo drew a lot of attention because they could usually be counted upon to do something entertaining, like hurl a tree trunk around their yard. Although Packy's notoriety as the first elephant born in forty-four years faded, he'd become renowned for his size. He was now the largest bull elephant in America, weighing nearly fourteen thousand pounds, the crown of his head reaching a height of twelve feet.

Pet, Hanako, Sung-Surin, and little Rose-Tu also had their share of admirers, but without a doubt, Belle was Queen. Delivering Packy all those years ago had guaranteed her celebrity status, but it was her palpably sweet nature that endeared her to visitors. She'd been described as possessing ladylike deportment, which Roger supposed was true, but she was so much more than that; she was kind, caring, humorous, stubborn.

"You don't love someone for one reason," he said in 2018, "but for all the things that make them up—even the stuff that drives you batshit."

Now that Belle's foot problems were public knowledge, she received wishes for a speedy recovery from admirers near and far. Get-well cards arrived in the mail every day, and secretaries in the front office reported an increase in telephone calls inquiring about Belle's condition. One reporter or another showed up almost daily looking for a scoop, but Roger had little patience for them since they interrupted his work. There was little to tell them, anyway.

The keepers knew surgery was the only available option if they hoped to save Belle's life. The question was how to do it. The first suggestion, that surgery be performed with Belle standing and under a local anesthetic, was rejected as being too risky for all concerned. The second route, considered the safest, was for her to lie down on command before receiving an immobilizing agent—a position Belle could no longer assume due to her frozen joint. Sedation of a standing elephant can be done, but it's ill-advised and dangerous because the animal goes into free-fall once the drug takes effect. Belle's elbow, already compromised by the old injury, could be seriously traumatized by such action. The third recommendation was to rig a sling around her, administer a sedative, and gently lower her to the floor with guy ropes, but the position in which her elbow would have to rest during surgery and the pressure of the elbow against the hard floor posed a risk of serious injury.

Her frozen joint wasn't the only issue to present an obstacle. The UC Davis veterinarians expressed concern over the leg's ability to adequately support her weight post-surgery. Increased stress on the elbow caused by shifting her weight to the right could lead to destabilization. Also, Belle was forty-five years old—not ancient by any means, but certainly no longer young. There'd been rapid turnover in her molars through the years and she was now on her sixth and last set. In the wild, loss of those final teeth would spell the end for her. That wasn't the case at the zoo, because they could find other ways to provide her with sustenance, but it was worth consideration.

After lengthy debate among keepers and veterinarians, and consultation with the professionals at UC Davis who had experience in similar operations, the decision was made to go forward with a sling and some type of cushion for Belle to lie on. They scheduled her surgery for mid-March of 1997, barely two weeks away. The first order of business became modifying an area of the elephant barn into a surgical suite. Roger wasn't sure the preparations would be ready in time, but everyone came together to help Belle.

Preparation for the event became an exercise in cooperation as several West Coast businesses stepped up and offered their services. Hoffman Construction Company coordinated materials and labor to transform an interior holding pen into the surgical suite. Kramer-Gehlen and Associates and Columbia Wire and Iron Works designed and manufactured parts of the suspension system. Allied Power Products provided the hoist. Charlie Anderson modified the design of his Anderson Equine Sling to fit Belle, and Care for Disabled Animals, Inc., manufactured it. Legacy Health Systems loaned a blood gas analyzer, and New World Manufacturing, Inc., constructed a special water bed designed to support Belle's weight and take pressure off her injured elbow. In a gesture of goodwill, the surgical team from UC Davis—esteemed animal specialist Murray E. Fowler, surgeons John Pascoe and Larry Galuppo, anesthesiologist Eugene Steffey, and surgical technician Rich Morgan—donated their time and services, while the Shiloh Inn provided them with free rooms.

Roger and his crew labored diligently to keep things running as close to normal for the elephants, but the animals were clearly curious about the noise and the presence of so many strangers. Newspapers published articles detailing the procedure. When reporters showed up at the barn, Roger told them to choose their questions carefully, because he was only going to answer them once.

In the days leading up to surgery, Belle's retinue of keepers worked to familiarize her with the equipment she would see, hear, and smell on the big day, as well as any part of the procedure they thought might spook her. The plan was to have the sling available several days before the actual surgery to give her time to acclimate to the sight and feel of it, but unforeseen last-minute adjustments made that impossible. The team improvised a makeshift sling out of old fire hoses and conditioned Belle to allow them to wrap the hoses around her body and legs. As her favorite among the keepers and the man she trusted most, Roger was paramount in this training, gently taking her step by step through the process, doing his utmost to not lose patience when she balked or fought the restraints. He was always aware of time passing too fast.

To accustom her tiny herd to their matriarch's absence during surgery and recovery, Belle was periodically separated from her constant companions Sung-Surin and Rose-Tu. The calf didn't like being kept apart from her devoted foster mother, but Sung-Surin stepped in and offered support.

As the clock ticked down, Roger kept his nerves at bay by focusing on details and preparations—anything to distract him from his fear of losing Belle. Thirty-six hours before surgery, her meals were withheld to remove any risk of aspirating the contents of her stomach into her lungs during the procedure. "She wasn't very happy with me about that," he recalled in 2015. "And it's not like I could explain it to her."

At the twelve-hour mark, water was denied. Roger stayed beside Belle the entire time, talking quietly, his gut a curdled mess of worry as he tried to reassure them both that everything would turn out just fine. A newspaper photo from that time shows them together, Roger with his hand out, Belle extending her trunk, a bit like Adam and God in Michelangelo's painting. For a moment, it holds them apart from the anxiety and bustle surrounding them. For the space of a few heartbeats, they are the only two creatures in the world.

At 7:30 a.m. on March 19, 1997, the seventeen-person surgical team mustered to review the procedure. Roger hung on the sidelines and listened intently as they outlined the plan. Belle would be sedated, secured in the sling, and anesthetized. The hoist would lift and lower her onto her right side on the water bed, and her left front foot placed on a platform for easy access. The operation was expected to last from two to four hours, depending on what the surgeons found when they opened the site. Afterward, the team would administer a drug to reverse the effects of the anesthetic and get Belle back on her feet as quickly as possible.

Roger and the barn team prepare Belle for surgery, securing the harness used to hoist her into the air and position her in the operating theater. (Larry Galuppo, MD, UC Davis School of Veterinary Medicine.)

By 8:30 a.m., they were ready to begin. Roger led Belle from her quarters into the surgical suite, where she received a soothing foot soak to soften the tissue. When that was finished, he and the other keepers—all familiar faces in an attempt to make the day seem routine—positioned the sling around her body and began to attach the straps to the hoist's metal framework. Belle shifted nervously and tossed her head. When she failed to respond to Roger's efforts to quiet her, she was given a sedative. Fifteen minutes later, she was relaxed enough for the team to resume securing the sling straps to the hoist, a process that took an additional twenty minutes.

"I was so proud of her I could have cried," Roger said, speaking in 2015. "Cooperating like she did, letting us do what we needed to do and trusting that we wouldn't do anything wrong, well, that was just Belle being Belle."

Long ropes were tied to each of her legs so they could be manipulated individually during the procedure if need be, and Belle was injected with a general anesthetic. When it took effect, the hoist engaged and the sling was raised. For a moment, the hoist became blocked, not quite lifting Belle clear of the floor, and the crew panicked, but representatives from the crane company, the waterbed manufacturer, and the sling company were standing by, and the problem was quickly resolved. In three minutes, Belle was fully suspended in the air.

The team worked quickly to unroll a floor mat and the empty water bed beneath her. In a feat of astonishing orchestration and coordination that took only one minute to complete, they lowered Belle onto her right side and elevated her left foot onto its platform. Two hoses pumped warm water into the water bed. Belle was intubated, started on gas anesthesia, and connected to equipment to monitor her heart, blood oxygen level, and respiration. When that was complete, the operation began.

As Belle snoozed, Roger stood at her back, one hand touching her at all times as he watched over the wide expanse of her belly. Never the squeamish sort, he followed the proceedings with interest. He could handle blood and guts, meat and bones; it was emotions he had trouble with, the realization that this was happening to someone he loved and no one could predict the outcome.

Over the next few hours, doctors Pascoe and Galuppo amputated the infected area. Visibility was poor due to hemorrhage, but they debrided deeper tissue as best they could. They placed bone cement mixed with antibiotics in the wound, which was left unsutured to heal by secondary intention. This route is taken when large wounds with considerable tissue loss make it impossible to bring the edges of the wound together. Healing takes longer, scarring is greater, and the chance of infection increases. When the surgery was complete, the entire foot was bandaged.

Within minutes of the anesthesia being reversed, the waterbed was empty and Belle was standing on her own. As keepers began to release the sling straps from the metal frame, she tossed her head violently, crying out and lunging. Roger did what he could to calm her, but her refusal (or inability) to listen made her a danger to herself and to the team. Withdrawing to a safe distance, they waited as another dose of sedative was administered. Once it took effect, they removed the sling and released her.

Belle's response to commands remained unreliable. For the next hour direct contact with her keepers, including Roger, was prohibited. When she could walk properly and be trusted to follow simple directions, they opened the doors to the surgical suite and brought Belle to her recovery area in the front exhibit room.

The post-surgical atmosphere in the barn was an odd mix of dissipating worry and giddy euphoria. Everyone on the surgical team felt optimistic about Belle's chances for a full recovery. Roger worked his way through the group, shaking hands and expressing his gratitude.

"After twenty-eight years of working with her, I was *supposed* to feel something," Roger said, speaking in an interview in 2018. "But the surgeons and the companies and the public and all those who supported this effort to save her, they didn't *have* to, you know what I'm saying?"

Dr. Galuppo described the event as "an incredible team of people and one exceptionally intelligent animal that wanted to live."

In the days following the surgery, Portlanders turned out to wish their favorite pachyderm a speedy recovery. Bouquets of flowers and thick bundles of get-well cards arrived at the zoo.

Belle's bandage was changed daily. She received regular injections of antibiotics and was kept in her sandals. With the assistance of zoo volunteers, the keepers established an around-the-clock watch to monitor her progress. Every few days, Dr. Galuppo would travel up from California to check the surgical site. Even when it was at its most painful, Belle never hesitated to lift her foot and place it on the tub where he could reach it. It wasn't long before she'd lift her foot into the air as soon as she saw him coming. Dr. Galuppo admired her obvious intelligence and felt awed by the incredible bond she shared with Roger. Regular contact between the men, and their devotion to Belle, quickly transformed them from colleagues into friends.

Two weeks post-surgery tissue healing had begun, but radiographs revealed further degradation in a bone previously operated on, as well as the presence of bacteria. Belle's foot swelled and edema spread into her leg. With Roger standing by her head to provide comfort, she was sedated, given local anesthesia, and the wound area and bone debrided once more. The following week, radiographs showed bacteria and a large area of dead bone tissue in the toe previously excised. The zoo veterinarians began intravenous antibiotics and removed over a quart and a half of watery fluid mixed with blood from the surgical site.

Roger remained with Belle every step of the way, keeping her company whenever his other duties didn't demand his attention. He slept on a hay bale or in a chair just outside her room when he couldn't bear to be apart from her at night. RoseMerrie understood and did what she could to help, offering moral support and meals which he ate but rarely tasted.

Belle's appetite declined. She grew listless. The flesh above her eyes sank close to the bone and her head hung heavily between her shoulders. When Roger coaxed her onto the truck scale used to weigh the elephants, he was horrified to discover she'd lost over four hundred pounds since the surgery.

On April 22, 1997 the surgical team from UC Davis returned to Portland for a second attempt at resolving Belle's issues. As the team gathered, preparing to reenact the scenes from a month before, Roger stepped outside for a quiet moment and a smoke. His heart felt almost too heavy to beat. He leaned against the barn's outer wall, the concrete warmed by the sun, and briefly closed his eyes.

He was on his third non-filtered cigarette when John Pascoe stepped out of the barn to get something from the UC Davis van. Seeing Roger, he approached and said in his Aussie drawl, "A little on the tense side, mate?"

Roger blew a cloud of smoke. "You could play a goddamn tune on me, John," he said.

Dr. Pascoe nodded. "I promise you, Roger, if it can be done, we're going to do it. If it can't, it won't be for want of everyone giving it their best."

It was the pep-talk Roger needed. Crushing the cigarette with the toe of his boot, he followed the doctor back inside and resumed his place beside Belle. The team repeated the same anesthetic procedure as before, and when Belle was supine, the surgeons set to work. Once more, Roger watched from beyond the curve of Belle's abdomen, one hand stroking her back, talking softly to her under his breath. Beyond these walls it was business as usual in the barn, but the surgical suite seemed strangely quiet except for the murmured voices of the surgical team and the soft hiss of oxygen.

The surgeons removed more bone. They avoided hemorrhage in the site by using an improved tourniquet, which enabled them to see that the infection had advanced deep along the tendons, progressing upward from Belle's foot into her leg.

Reliving that awful day during an interview in 2015, Roger said, "None of us realized just how bad it was until John slammed his scalpel down onto the floor. That told me more than any amount of words ever would. That's when I knew it was over." He couldn't afford to let it show, but inside he wanted to scream, to die. "I would have bartered my life for hers, done anything to achieve a miracle."

He pressed his palm flat against her side, watched it rise and fall with her steady breathing, and felt the beat of her mighty heart.

Dr. Galuppo peeled off his surgical gloves and dropped them on the floor. Meeting Roger's eyes, he explained to everyone present what was happening inside Belle's leg. The infection had severely compromised the support structure of her foot. "To continue would be unfair to Belle," he said quietly. There was only one last kindness they could give her.

Belle was euthanized while under anesthesia. Outside in the yard, under a drizzling sky, Sung-Surin and Rose-Tu began to trumpet and bang on the door as the matriarch's heart slowed and stopped beating. Dr. Galuppo appeared startled, but Roger wasn't surprised. He understood the sudden vacancy they were feeling, the empty space where Belle's voice had resided for so long. His heart felt the same way.

Two hours later, once all of the equipment had been taken down and carted away, Roger opened the door to the operating theater and allowed Belle's herd inside to view her body so they could understand she was gone. Sung-Surin refused to cross the threshold, but little Rose-Tu bravely entered and walked around Belle, nudging her with her feet and head, stroking her with her trunk, uttering sad little cries and rumbles. She was pleading with Belle as Roger wanted to plead: Get up. Be alive. Don't go.

Eventually, Rose-Tu rejoined Sung-Surin, and they walked away together, a herd of two.

Suiting up in protective gear prior to dismembering Belle was the most surreal moment of Roger's life. His stomach churned with bile and stale coffee. He was grateful he hadn't eaten in a while, because he wouldn't have been able to keep it down. Back turned, he felt the eyes of the other keepers on him as they, too, prepared for the heinous task. He knew everyone would understand if he opted out and let them handle it. There'd be no recrimination. But he'd been with Belle in life and escorted her to the gates of death; he refused to drop the ball now.

When Roger looked up, he was surprised to find Jay Haight pulling on a hazmat suit. Several years earlier, his partner had lost the ability to tolerate putting down and cutting up old friends and had moved from the elephant barn to the African section of the zoo.

"Goddammit, Jay," Roger said. "What are you doing here? I thought you had enough sense to get out of this place."

Haight shrugged and zipped his suit. "It's like you always said. You don't let old friends go out alone." Side by side, they and the other keepers fired up their chain saws.

***

In the days immediately following Belle's death, the Oregon Zoo was inundated with telegrams, telephone calls, consolatory letters, and sympathy cards. A dozen roses left at the front gate bore a note thanking the zookeepers for taking such good care of Belle. Another note, accompanying a bouquet of white azaleas, read: "Belle—we'll miss you terribly."

The children in Peggy Dale's first grade class at Sacajawea Elementary in Vancouver, Washington sent Roger a thick packet containing drawings of Belle. The children shared their memories of her and expressed sorrow for Roger's loss in messages written in crayon. Roger still has those letters, safely tucked away in a box of memorabilia.

Director Sheng sent Roger a letter expressing her appreciation for his "leadership in holding people together, in being committed to solving problems, and in always keeping up the spirit…You are a fine elephant manager." A card arrived from Michael Burton, executive officer at Metro: "It is an uplifting testament to the human spirit to note the efforts, sympathy and grief shown by so many people on behalf of Belle. And there is no real way to measure the depth of one endeavor by those who assisted in providing the intensive care for her. Please let this note serve as an expression of thanks from all the people in this region for the personal care and concern you gave during a very difficult time. Your courage and professionalism are well noted; your passion and commitment will be well kept."

Roger received condolences from zoo directors and other elephant handlers across the country. He heard from Gunther Gebel-Williams, from Democratic Senator Bob Packwood, and from Republican Senator Mark Hatfield, who'd often brought his children to the zoo and never failed to thank Roger for caring for his party's symbol, to which Roger, a lifelong Democrat, always replied, "No offense, Senator, but these are nonpartisan elephants." Grandparents wrote about seeing her perform when they were children, back in the days of the yard shows, or recalled visiting her after Packy was born. Kindergartners sent drawings of Belle surrounded by flowers and lopsided hearts, struggling to express what might be their first experience with death.

Roger read every card and letter, and was deeply appreciative—but not at all surprised—by how many people Belle had touched over the years. What stunned him was the concern directed toward *him* by hundreds, if not thousands, of total strangers. Wherever he went, around the zoo or in town, they stopped him to offer condolences and their thanks for all he'd done to help her.

To Roger, all of the sympathy seemed wrong somehow. "I more than half expected to be stoned for letting Belle die," he said repeatedly during interviews in 2015. Instead, he was swept up in an immense wave of love and support that often closed his throat with emotion.

On April 24, 1997, the zoo held a service outside the elephant barn to honor Belle's memory. Five hundred mourners bearing cards and flowers arrived to pay their respects and tears flowed freely. Director Sheng wept as she spoke of bringing a banana to Belle shortly before her surgery, remembering how the elephant had breathed gently on her. "I think forever she stole my heart," Sheng said.

The last thing in the world Roger wanted was to speak at that gathering. He preferred to stand among the shadows inside the barn and listen until he could take no more, then return to tend the remaining elephants. But his participation was expected. People had come to share their grief, but also to hear from *him*: the salty-tongued elephant keeper whom many had grown up seeing during visits to the zoo. Roger was the man thousands had read about in zoo-related newspaper articles over the years. Everyone admired his devotion, and they were all there to lend their support.

Roger stepped up to the microphone as though facing a firing squad. In his uniform, campaign hat, and rubber boots, he looked out across the crowd and wondered what to say, where to begin, how to convey what Belle had meant to him. He thought of her stubbornness when dealing with the herd, and her refusal to back down to any challenger even after her injury. He remembered her sense of humor and that glint in her eye when she'd trick him into opening the office door by making the knock-knock noise with her trunk. He thought of the gentle awareness of her presence that followed him around the barn even when he wasn't with her, as if part of her spirit always walked beside him. He recalled how much she'd trusted

Roger at Belle's memorial service, in the Zoological Gardens. (*The Oregonian*, 1997)

humans no matter what they asked of her. He smiled inwardly at all the times she'd stolen calves from their mothers, confident that she was the only one who knew how to properly care for and protect them. He remembered the day she rescued him from Hanako, his pride in her as she'd cooperated during the difficult days leading up to surgery, and how she seemed to understand that they were trying to help her. Belle's son may have received the elephant's share of attention, but she was the real star.

Somewhere in the midst of all those memories, he began to speak. He shared some of those thoughts, but not all of them. Some things were simply too raw, too painful, to bring out into the open. When he finished, he stepped down from the podium and slowly made his way through the crowd, shaking hands as he headed for the barn. As always, there was work to be done.

***

After Belle died, life at the zoo felt wrong to Roger. A piece of him was missing, leaving a hole he couldn't patch with work or family. Each elephant's death had stolen something from him, but Belle's was the hardest to get past. Unwilling to burden others with his feelings, he didn't speak of it, but he understood that he'd hit some sort of wall. He wasn't certain whether he could recover, or if he even wanted to.

Roger was angry because he wasn't magic. Belle saved his life but he hadn't saved hers. That the problem had been beyond his power to solve was no consolation. He believed if he'd only been smarter, worked harder, or threatened more people, things would be different and she'd still be waiting for him every morning.

In a slightly macabre way, she was. Unbeknownst to anyone except Haight and one or two particularly close friends, Roger had taken Belle's skull and buried it deep in the manure bunker where no one would find it. It was a spontaneous decision, sparked by a nebulous desire to pay homage to her life.

The skull lay hidden for months while insects and microbes ate away the soft tissue. Every so often, when no one else was around, Roger would dig it out and hose it off to see how things were coming along. A strange, psychic bandage seemed to surround him at those times, as if something bolstered his soul against the well of grief deep inside.

By autumn of 1997, the skull was pristine, picked clean and white. One afternoon before going home, he dug it out, washed it a final time, bagged it, and placed it in the bed of his truck. The next day, he and his brother, two nephews, and a couple of friends loaded up their guns, camping gear, and boxes of supplies into several vehicles and headed into the mountains for their annual hunting retreat. Roger drove his old pickup, the interior redolent with elephant musk. If he seemed quieter than usual, no one remarked on it.

When they arrived at their site, all hands helped to unpack the vehicles, set up the cabin tents, and start a fire going in the woodstove. They secured their food out of reach of marauding bears, unrolled their sleeping bags, and opened the first of the cold beers as they settled in.

Roger kept himself busy and distracted, but the back of his mind was focused unerringly on the task that lay ahead. As the sun began to descend, washing the sky with color, and the air turned cool, he carried a fresh bottle of rye and the bag containing Belle's skull to a nearby clearing where there was a big blow-down of dead trees. His friends and family watched him go, looked at each other, and said nothing.

With ceremonial intent, Roger used the big knife on his belt to hew limbs into foot-long sections, losing himself in the methodic *whack* of blade against wood. The sky darkened, bleeding from red and orange to purple, from there to deepest blue, and then black. Stars emerged, shining cold light down on the man engrossed in his work. Still Roger cut, working up a sheen of sweat which cooled in the breeze stroking his arms and the back of his neck.

When he'd finally chopped enough wood to suit him, he knelt to clear an area of grass. Placing Belle's skull in the center, he looked at it, unable to move, arrested by the sight of all that remained of his dear friend.

Now the tears came. Roger trembled and remembered Belle, all she'd been to so many and all she'd meant to him. One angry handful at a time, he stuffed the skull with wood and tinder, then set a match to it. Dry as the fuel was, the flames caught and spread rapidly. Fire danced where Belle's bright eyes had been, winking and flickering, bringing false life to those hollow caverns.

Cracking the seal on the bottle of rye, Roger sat back and methodically fed log after log into the fire. As the flames slowly consumed the skull, he drank, wept, and cussed. Leaning back on one arm, he tipped his face to the sky, gazing at the stars through a haze of smoke and tears.

Chapter *Fourteen*

MOVING ON

1997–2017

After the ceremony at the hunting camp Roger returned to Portland, focused on family and work, and tried not to think about the emptiness that filled him. Some moments of happiness shone through the despair; he welcomed another grandchild, Michelle's second daughter, Gabrielle, whom Roger babysat with the same devotion he gave to her older sister. He grew his yearly crop of tomatoes, cucumbers, and garlic, and generally did what he could to get on with life.

Despite Haight's efforts, Hanako continued to present a threat to keeper safety. On December 9, 1997, she moved to Point Defiance Zoo, which managed elephants solely under a protected contact protocol and was well-equipped to handle difficult animals. Hanako joined matriarch Cindy and another dangerous cow named Suki to become a herd of three. Although Hanako was nervous about the change and much subdued afterward, the Point Defiance elephant keepers fell in love with her immediately and worked diligently to help her regain her confidence. For her part, Hanako did her best to please. Her victories came in small steps rather than bounds, but over time she settled in and became comfortable with her new surroundings.

Back at the Oregon Zoo, Roger cared for the remaining elephants—Pet, Packy, Sung-Surin, Rama, Hugo, and Rose-Tu—with the same diligence and conscientious devotion as always, but his heart wasn't in it to the same degree. Maybe it was age, he told RoseMerrie one night over dinner, or maybe it was the combined loss of his parents and so many elephants in such a short time, but he didn't feel the same fire as before. The barn seemed vast and echoing with only six animals in residence. He had no energy for confrontations with administration and couldn't bear the thought of losing even one more animal. Emotionally and physically wrung out, he thought maybe it was time to retire.

RoseMerrie cautioned him to act slowly. She said his feelings were likely due to Belle's death and he might feel differently once he had time to grieve. It would be a shame to leave a job he loved before he truly was ready to give it up. He was also sixty now, with only five years until retirement. Financially speaking, it might be worth the wait if he could manage.

Her advice made sense, so Roger didn't tell her he was looking for an excuse to leave. Instead, he agreed to wait and see how the coming months played out. Shortly before Christmas 1997 they consulted a financial advisor and discovered an early retirement would actually leave them better off. Roger decided the time had come. He drove to the zoo before he could change his mind and gave notice of his plans to retire at the end of February.

The news hit the zoo community hard, upsetting coworkers in every department. Roger was a fixture at the zoo. The younger keepers who'd worked with him had never known another senior elephant keeper. They couldn't imagine anyone, no matter how experienced, filling his muddy, dung-spattered boots.

On his sixty-first birthday, February 28, 1998, Roger took his final walk through the barn. He'd have liked to do it alone, but of course there were reporters and well-wishers. A picture taken that day shows him clad in his barn coat and familiar campaign hat, his union button polished and prominently displayed. Behind him, Pet and Rose-Tu reach their trunks past the bars, no doubt sensing his sorrow. His hand rests on Pet's trunk, and her eye peers over the brim of his hat.

Behind the darkened lenses of his glasses, Roger was in tears. After shaking the last hand and heading for the door, he made an abrupt turn and nailed one of his old campaign hats to the filing cabinet as a parting shot at the universe.

***

Roger's first morning without the job felt strange and a little intimidating. He wondered how to fill the long hours in each day without animals to shift, manure to shovel, and hay to haul. More than that, he wondered who he was now that he was no longer an elephant keeper.

He decided he could focus on being a husband. He'd neglected RoseMerrie too often in the past and intended to make up for it in the time he had now. He took over the laundry, freeing his wife from that duty, and threw himself into caring for their yard and garden more devotedly than ever before. RoseMerrie provided the quintessential retiree "Honey Do" list, but Roger plowed through it in a few weeks. With little to occupy him, his thoughts wandered back to his elephants, the smells of the barn, and the thousand-and-one little surprises the day could bring.

Jay Haight and Charlie Rutkowski kept him updated on what was happening at the zoo, but hearing the news secondhand wasn't the same as being there. Like Tucker before him, what Roger felt most acutely was not the absence of the

Pet and Rose-Tu reach out to Roger as he makes his last walk through the barn as their keeper. (*The Oregonian*, 1998)

elephants—although he certainly missed them—but that of his fellow keepers, the band of brothers and sisters with whom he'd bonded through the years of shared experience and love of the animals.

In an attempt to keep his hand in the game, he signed up to volunteer in the elephant barn on an as-needed basis but was never called. That was probably just as well; Roger knew he would never be able to keep silent if he disagreed with anything about the animals' care. He didn't want to become the old-timer who got underfoot and caused problems.

That's what he told himself, anyway. But there were deeper reasons why he kept his distance, reasons that prevented him from accompanying RoseMerrie whenever she took their granddaughters to see the elephants. Each animal death, each conflict with management that turned out badly, and each incident when Roger felt he'd fallen short became a brick in the wall he'd unwittingly built around himself. The more he stewed over the past, unable to view it with a kinder perspective, the higher and more impenetrable the wall became, until he couldn't even see past the failures to all the good times and the many successes. Embittered beyond all reason, unable to allow himself those fond memories because he felt he didn't deserve them, Roger labored to push every memory of the zoo to the back of his mind and forget it entirely. What he sought was oblivion. His inability to find it was just another shortcoming he suffered in silence.

To make matters worse, RoseMerrie's breast cancer returned in the spring of 1998. Surgery alone couldn't defeat it this time, so she started chemotherapy and radiation. The long recovery period gave her plenty of time to think about the future and how she wanted to spend it. She retired in the spring of 1999, and the two of them fulfilled a decades-long dream of leaving Portland's damp climate for the drier air of eastern Oregon's high desert country.

Living three hours from the zoo instead of twenty minutes gave Roger an even better excuse to stay away. On occasions when they returned to Portland to visit family, RoseMerrie never failed to take their granddaughters to the zoo, but he always found an excuse to avoid going despite their entreaties. RoseMerrie had lived with his depression for over thirty years and was long past being tolerant. She sympathized with his pain but not his determination to do nothing about it. It was one of the few things they clashed over.

Every so often, someone who learned of Roger's background would suggest he write a memoir. "Who the hell wants to read about some old shit-shoveler?" he'd say and change the subject. Michelle bought him a handheld tape recorder in hopes it might ease the process, but the machine remained in its box. Offers to teach him how to use a computer were met with a flat glare that brought a swift end to the discussion.

RoseMerrie's breast cancer returned again in 2000, now metastasized to her lymph nodes. She decided she'd had enough of this nonsense and opted for a double mastectomy. While she remained in the hospital, Roger shuttled back and forth in the old truck—which still smelled of elephants—and worried despite her reassurance she felt fine and would soon be home. At night, alone in their bed and unable to sleep, he tortured himself with thoughts of how impossible life would become should she die. After she came home, he hovered solicitously, asking repeatedly how she felt or if there was anything he could do. She made a full recovery, and they danced together at daughter Melissa's wedding in 2005.

RoseMerrie might grouse about her husband's depression, but she also understood him better than anyone. She knew he carried in his heart a longing for the elephants. She couldn't bring back Belle, but whenever a newspaper article appeared about the zoo, particularly the elephants, she'd point it out to Roger or leave it prominently displayed where he couldn't help but see. After he read it, she'd carefully clip it out and store it in one of several plastic bins of memorabilia. Over time, the clippings stacked up, announcing such things as the arrival of Borneo pygmy elephant Chendra, Hugo's death in 2003, Packy's annual birthday fete, and the 2005 acquisition of a new bull named Tusko.

One August afternoon in 2006, as Roger sat on the back deck enjoying a glass of iced tea after a day of mulching tomatoes in the sweltering heat, he received a devastating telephone call from zoo veterinarian Mitch Finnegan. Pet, his old tobacco-chewing pal and the sole surviving member of the zoo's original herd, was going to be put down.

Roger closed his eyes and let Finnegan's words wash over him. God knew how many hours he'd spent on Pet's feet, coaxing her to let him work, battling the degenerative arthritis that plagued her for years. Lately, she'd been observed using her trunk as a crutch and leaning against the bars of her cage to ease the weight off her feet. The familiar protocol of exercise, anti-inflammatories, and pain medication was no longer working, and the veterinary team had run out of options.

Finnegan offered his condolences and his regret at being the bearer of bad news. Roger assured him it was all right and thanked him for the kindness. "A call from you is better than the shock of hearing some stranger announce it on the news," he said.

There was a pause on the other end, and then Finnegan said, "I can sneak you into the barn to give her a last pinch of tobacco and say goodbye if you want."

Roger dug a bandana out of his back pocket and swabbed his eyes. "I appreciate that, Mitch, but I'm only now past the nightmares of Tuy Hoa and Rosy and Belle. I really don't think I got it in me."

When the announcement of Pet's death hit the newspaper two days later, he felt it like a physical blow. He threw the article in the trash, but RoseMerrie later dug it out to store with the rest.

Roger was torn, drawn to any information about the elephants yet wanting to escape the past. He couldn't do both, and couldn't choose one over the other. He pretended not to care, but those who loved him knew better.

Toward the end of that year, news of Rose-Tu's first pregnancy was shouted from the Portland rooftops. The announcement brought home to Roger how many years had passed since his final journey through the barn. She'd been a youngster then, not yet four years old and a long way from breeding. Now she was set to become a mother.

In August 2008, Rose-Tu delivered a 286-pound son, Samudra (Hindi for "ocean"), who carried the distinction of being the first third-generation elephant born in the United States. In time, he also proved to be a "true tusker," an increasingly rare genetic trait in male Asian elephants as more are killed for their ivory and fewer with the tusk gene survive to pass it on. Samudra's father, Tusko, lost both of his tusks to chronic infection.

Rather than nurture her child, Rose-Tu appeared frightened of Samudra and immediately tried to trample him. Keepers stepped in to separate them before the calf could be injured. Over the next several days, they worked to safely reintroduce the pair. Rose-Tu soon came around and quickly transformed into an exemplary mother, just like her mother Me-Tu and grandmother Rosy.

Roger wasn't surprised when he read about her initial reaction to the calf. Fourteen years had passed since Rose-Tu's birth. She'd never smelled a pregnant cow, witnessed a delivery, served as an auntie, or had an older cow to inform her. Twenty-six-year-old Sung-Surin had some experience with calves—she was four when Chang Dee was born, and twelve when Rose-Tu came along—but she'd never been pregnant or displayed any interest in breeding. All she had to offer Rose-Tu was moral support, and that just wasn't enough.

In 2011, Roger and RoseMerrie welcomed a grandson, Silas. Meanwhile, in Portland, the zoo announced plans to overhaul the elephant compound, increasing it to six acres with a projected completion date at the end of 2015 and a budget of $59 million. The new facility, to be called "Elephant Lands," was designed to improve the lives of the animals by offering such amenities as a larger area to roam, a full-immersion swimming area, and remote feeding stations. The secondary goal of the facility was to educate the public about the risk of extinction and the work being done by the Oregon Zoo and other organizations around the globe to guarantee the elephants' survival.

Roger swore when he first read about the project. He might have achieved miracles with that much space and money to work with. It was hard not to feel jealous.

In 2012 Rose-Tu delivered a second calf, a female named Lily. This time, she stepped easily into her role as mother. As the months passed, Roger avidly followed stories about the calf and all reports about the visionary habitat. A flicker of his old enthusiasm returned. He no longer needed RoseMerrie to point out articles and clip them; he did that himself, even with those that brought sadness.

In 2013, an outbreak of tuberculosis started among Packy, Rama, and Tusko. The bulls were isolated from the herd, started on a months-long treatment regimen, and kept at least one hundred feet from the public. Seven staff members who had close contact with the elephants developed a latent form of the disease, noninfectious and without symptoms, and recovered fully.

Although Rama responded well to the tuberculosis treatment, his old leg injury, the greenstick fracture that never properly healed, caught up with him at last. On March 30, 2015, Rama was euthanized at the age of thirty-three.

A few months later, Roger watched live on television as the old barn was demolished. All that he'd known so intimately—the keeper alley and front room, the interior holding pens and his office, the hay storage and yards, even the room where Belle took her last breath—was reduced to a pile of masonry rubble, dust, and thousands of memories. He knew progress demanded such things, but it still hurt to see it go.

On December 16, 2015 the newly created Elephant Lands opened to the public. Chendra, Tusko, Samudra, Lily, Sung-Surin, Rose-Tu, and routine-loving Packy had slowly been introduced to the exhibit over the prior months as one area after another reached completion. Now it was theirs to explore and enjoy in its entirety, watched over by their keepers and visitors.

Roger desperately wanted to see it for himself, but dared not go. Too much time had elapsed. He worried no one he knew would be there, but also feared running into someone familiar and not knowing what to say after refusing to return for so many years. He yearned to visit his old friends Packy, Sung-Surin, and Rose-Tu, and simultaneously dreaded the changes he would see in them, a reminder they too were mortal.

Then the unthinkable occurred. Packy—the elephant who'd put Portland on the map, Tucker's hairy little darling, the first calf Roger ever laid eyes on and his performance partner in the yard—was euthanized in February 2017 due to complications and drug-resistance stemming from his tuberculosis. The announcement of his death, delivered with gentle kindness in a phone call from Mitch Finnegan, swept Roger's legs out from under him and sent him spiraling into the deepest depression he'd suffered in a long while. He'd never said so out loud, because someone in his family was sure to hold him to it, but he'd hoped to see the old boy one last time. Now that dream was lost forever.

Chapter *Fifteen*

FULL CIRCLE

March 2017

For nineteen years, Roger tried to lock away his memories of the zoo, but every newspaper article or television broadcast brought them rushing back in vivid color. The elephants haunted his waking hours and paraded through his dreams as he slept, refusing to be forgotten. Packy was lost to him, but his old friends Sung-Surin and Rose-Tu remained. While there was still time, he wanted to witness for himself that they were healthy and happy.

Sensing Roger's growing desire, RoseMerrie, Michelle, and Melissa cajoled him to visit the zoo, something they hadn't done in years. Roger resisted at first out of habit, but soon began to consider that a visit might actually be possible.

On March 30, wearing the same campaign hat he'd worn on his last day in the barn, Roger led his family through the front gate of the Oregon Zoo and shook hands with elephant curator Bob Lee, who was waiting to meet them. Within moments, the two elephant men were chatting like old comrades. They'd never worked together, for Lee had started in the barn right as

Roger retired, but Lee knew him by reputation and the stories about Roger still circulating around the zoo.

Lee began their day with a quick tour of the grounds so Roger could see some of the changes made since he'd been gone. Lee had arranged for a multi-passenger golf cart to ferry his visitors while he walked alongside providing commentary and explanation. Roger thanked him, but insisted he was up for the walk. For as much as he'd dragged his feet when it came to visiting the zoo, now he was suddenly energized, talkative, and seemed decades younger than his eighty years. For all his fear of confronting the past, he now strode eagerly into the future.

As they came around the first bend, Roger discovered a surprise waiting for him: a group of former coworkers eager to welcome him back. Despite the passage of years, he had no difficulty placing names to the faces of these women and men. He remembered many as teenage volunteers and young adults just beginning their zoo careers; now they were parents of children graduating high school. Roger shook hands, humbled and honored to be so well-remembered. The group laughed as they rehashed old stories. Keeper Diana Bratton recalled the time she nearly let a bull elephant into a room where Roger was standing.

"Never did it again, though, did you?" Roger said with a smile.

Veterinarian Mitch Finnegan watched the proceedings with quiet pleasure. "Roger is one in a million and I don't think we'll see his kind again," he said. "Our loss."

Roger and his retinue followed Lee along the trail into the Great Northwest exhibit to gaze at black bears, California condors, bobcats, river otters, bald eagles, and many other animals native to the region. Lee explained that thanks to the zoo's efforts, the California condor had recovered from an unsustainable wild population of just twenty-two birds to more than four hundred, over half of which flew free in the wild.

Roger turned in a slow circle, trying to take it all in. He gestured toward the northern sprawl of the exhibit. "This was nothing but woods in my day," he said. "They always talked about expanding, but there was never enough money." He shook his head in quiet astonishment. "If you blindfolded me and set me down in the middle of all this without telling me where I was, I'd never be able to guess. This is wonderful."

The words echoed inside him. Despite his fears, coming here had been a good idea.

They emerged from the exhibit and followed a walkway aptly named "Zoo Street," which took them past the carousel and polar bears, the North Meadow and Primate Forest, until they reached the new Veterinary Medical Center. In addition to meeting the U.S. Green Building Council LEED (Leadership in Engineering and Environmental Design) Gold Certification Standards for Sustainability, the facility contained a 30,000-gallon rooftop cistern to harvest rainwater, ambient light sensors to adjust interior illumination, solar-heated tap water, surgical suites with tables strong enough to accommodate an adult zebra, heated and rubberized floors, climate-controlled patient areas adaptable to any species, rolling skylights to provide fresh air and a view, and many other features.

"What a vast improvement over the old place," Roger said to Mitch Finnegan. "It must be a joy to work here." They paused near the exit, where another small crowd of Roger's fans had gathered. One woman asked for his autograph—something he never in his life expected to happen—and handed him a copy of a research paper in which he'd participated so many years ago he barely remembered it.

"Are you sure you want me to deface this?" he asked her. When she assured him she did, he scrawled a shaky signature and shook his head in wry bemusement at those waiting patiently for a moment of his time. "Don't you people have better things to do with your lives?"

Leaving behind his cadre of old friends and new fans, Roger and his family followed Lee outdoors into the rare spring sunshine. Spying movement in a nearby exhibit, he glanced over and stopped dead, eyes riveted on the young elephant strolling unconcernedly in their direction.

Roger's stillness was that of the discouraged seeker who against all odds finally encounters the treasure he most desires. His longing was a palpable tug they all felt; an invisible cord drawing him forward, a vibration in the air like infrasound. Not for the first time, he reminded his wife and daughters of the elephants he so deeply loved and had so diligently cared for. Without a word, he headed toward the elephant barn. There was nothing the rest could do but follow.

When compared to the complex where Roger spent thirty years of his life, the six-acre Elephant Lands habitat seemed massive. Gone were the cages and concrete of old. Now the elephants enjoyed free-roaming access to indoor and outdoor areas, hillsides, mud wallows, and a 160,000-gallon self-cleaning swimming pool large enough to accommodate the entire herd and deep enough to allow full submersion. Hard-packed soil had been replaced by a type of round sand that won't compact, which keeps the ground soft and allows rain to wash through into drainage rather than collect on top where it creates the damp environment so detrimental to the elephants' feet. State-of-the-art ventilation and heating systems adapted to the region's varied climate provided coolness on the hottest days of summer and warmth when things turned frigid. Timed feeders dispensed food at random intervals to keep the elephants on the move, enabling them to forage much like their wild cousins.

Like the veterinary medical center, Elephant Lands also meets the LEED Gold Certification. As part of the Oregon Zoo's ongoing commitment to a green initiative, heat diverted from the cooling pools in the polar bear exhibit passes through geothermal coiled pipes to warm the elephant exhibit. Rainwater from the roof of Forest Hall collects in an underground cistern and is reused to clean the facility and flush toilets. Solar power provides some electricity and also heats water for elephant baths. Weather permitting, fan power is cut to utilize natural ventilation.

"Talking to Roger about his time with elephants was like listening to stories about space exploration from John Glenn or Neil Armstrong," Lee said later. "He experienced a life with elephants that few people had the opportunity to do, and no one will ever be able to do again."

Lee explained that Roger's extraordinarily devoted work ethic had inspired generations of elephant care professionals to observe their animals closely, listen to them, and learn from them; to work with them rather than force their compliance, and to fight in the elephants' best interests. "Roger's legacy is that elephants in zoos across the country and throughout the world are living more natural lives. They're in family groups, raising young and being given the opportunity to [move about as they choose]."

Some of the best elephant care programs in the world were clearly inspired by the direction taken by Roger and his keepers so many years ago. Their work created a path leading to a better understanding of the animals they hoped to preserve for generations.

There's no single right answer to the question of elephant conservation. Some solutions that have been suggested and implemented include stronger anti-poaching practices and a worldwide ban on the sale of ivory; redesigning zoos to focus on conservation and breeding; the increased use of artificial insemination; and the establishment of sanctuaries where elephants can roam and breed at will, unmolested and rarely visited.

To decrease conflict with humans in the elephants' native habitats, conservationists have begun using trained elephants to scare their wild cousins from villages and farms; managing land to increase elephant territory and re-establish migratory routes; and using chemical signals and bees, which elephants apparently detest, to deter crop raiding and keep elephants away from certain areas. Conservationists have also worked to introduce viable alternative economic opportunities for those who depend upon the elephant for their livelihood, and begun educating indigenous peoples in the vital role elephants play in their ecosystem.

Closer to home, the Oregon Zoo's Future for Wildlife program partners with the Oregon Zoo Foundation (OZF) to provide grant money to local and global conservation efforts. A big proponent of individual action, the zoo encourages the public to explore how their everyday choices affect the environment. The Oregon Zoo also supports a broad range of efforts to help wild elephant populations through the work of the International Elephant Foundation and other field conservation programs. OZF earmarked $1 million as a permanent self-sustaining source of funds to support Asian elephant conservation, making it the Foundation's fifth $1 million endowment dedicated to conservation, education, and animal welfare.

After the tour of Elephant Lands, Lee led Roger and his family to a fenced area for some one-on-one communion with the elephants. This was what Roger had been waiting for—hoping for—but had not dared to expect. Lee had already done so much.

The first elephant to arrive was Rose-Tu, now a mature twenty-two-year-old.

"She looks just like her grandmother, Rosy," Roger said, his voice filled with longing. He'd often wondered if the tragic loss of Rose-Tu's mother followed so closely by the death of Belle had turned her mean-spirited and difficult, but now he had his answer. Healthy, energetic, bright-eyed, and a mother twice over, Rose-Tu eyed him with curiosity, tail switching, and pressed against the bars separating them.

"Do you think she recognizes you?" asked RoseMerrie.

Roger shook his head. "I don't know." It certainly seemed like she might.

Lee ordered Rose-Tu to step away from the bars. When she ignored him—"Just like her mother," Roger said—Lee redirected her by initiating a routine of training maneuvers, popping a treat into her mouth every time she complied. One after the other, she offered her best trunk up, foot lift, head shake, open mouth, and sit up. Convinced she would now pay attention to commands, Lee motioned Roger forward and handed him a banana. As he slipped the fruit into Rose-Tu's mouth, RoseMerrie wiped away a tear.

Rose-Tu squealed her thanks and begged for more. Roger obliged with a cantaloupe, which she popped like a grape and swallowed without chewing.

The rest of the herd appeared, eager for their share of the treats. Four-year-old Lily arrived first, then her big brother, eight-year-old Samudra, followed by Chendra, the sole Borneo pygmy elephant in the United States.

Roger watched them, his throat tight with emotion as Rose-Tu caressed the contours of Lily's face with her trunk the same way Me-Tu had touched hers, and Rosy had touched Me-Tu's. Seeing them together, witnessing Rosy reborn in her great-granddaughter, Roger felt uncommonly proud of what he'd done to help make this day possible.

He cleared his throat. "That Lily is beautiful." He gestured at Samudra. "He's going to be a magnificent bull. And Chendra is cute in a plain-as-mud sort of way." He shook his head in amusement. "You can tell she's sweet-natured and happy."

Sung-Surin was last to arrive, as if this grande dame of the zoo understood the significance of this occasion and wanted to make an entrance in keeping with her status as matriarch. When Roger last saw her, she was a gangly teenager coping with the loss of Belle and doing her best to be a foster mother to Rose-Tu. Now she was a tall, self-possessed thirty-four-year-old cow, an experienced midwife and auntie.

Roger's eyes grew misty at the sight of her. "She looks great," he murmured, his voice husky with emotion. Head tipped to one side, he appraised her with an ability honed over thirty years. "She's a quarter, no, maybe a third bigger than Pet was, and she's better built, not as pigeon-toed." He shook his head. Seeing her made him wish that he'd taken Finnegan up on his offer to feed Pet tobacco one last time before she died.

He looked at Lee. "You know, I'd really hoped to see Packy."

The curator's face fell. "We tried to keep him comfortable, keep him going, but in the end we just couldn't."

Roger reconnects with Rose-Tu during his visit to Oregon Zoo's new space, Elephant Lands. (Photo by Melissa Crandall, 2018)

"No, you did the right thing," Roger said. "If you're going to work this field, you have to be responsible for everything, and that includes knowing when to let them go."

On orders from keeper Gilbert Gomez, the elephants shuffled into a line along the fence, eager to be admired and pampered. Roger watched proudly as his five-year-old grandson stepped forward at Lee's invitation to feed Rose-Tu. Smiling broadly, Silas offered one banana after another to the elephant's trunk as keeper Pam Starkey distracted the younger animals with treats.

Roger leaned toward RoseMerrie. "What astounds me is that Si didn't hesitate to approach," he murmured. "He just went right up to her."

She smiled. "I guess that means he's a born elephant man."

Roger beamed.

While Silas, Michelle, and Melissa fed the elephants, Lee brought Rose-Tu into a more secluded area for a private visit with Roger. He stood quietly before her, hands folded together, keeping a respectful distance, poignantly aware she was no longer his.

"You can touch her," Lee said softly. Roger glanced at him, saw the truth in the curator's eyes, and stepped to the fence. Reaching between the braided strands of wire, he laid his gnarled hand flat against the wide base of her trunk.

Rose-Tu and her former keeper regarded each other. Something flickered in the depths of her eyes, and with a subtle shifting of feet, she leaned toward the old man as if wishing the barrier between them could evaporate and let her get closer. Clearly, so did Roger.

Lee offered him another banana. Eyes locked on Roger's face, Rose-Tu gently removed it from his hand with her trunk and conveyed it to her mouth. "She is so incredibly smart," Lee said. "She learns fast."

Roger nodded. So had her mother. So had her father Hugo, for that matter; he just had an issue with complying. With parents like that, Rose-Tu always had a lot going for her. "She's a gorgeous girl," he said, and smiled at Lee. "I wouldn't trade her for a painted pony or a speckled pup."

Lee laughed. "Me, neither."

Roger accepted more fruit from Lee and handed it to Rose-Tu. "When I got in the business, there were about 100,000 Asian elephants left and almost five million Africans," he said. "Now look where we are. I saw a blurb in the paper that Japan had agreed to suspend trade in ivory for one year. We're talking 620 to 640 days of gestation, followed by decades to grow those tusks. How much difference is a one-year suspension going to make?" He shook his head. "I used to think elephants in zoos were a good idea and then I spent thirty years with them. Now I just don't know. They can't compete with us. Any animal that's tried has failed. But if not zoos, then where?"

Roger let himself be pried away from Rose-Tu by the promise of a tour through the main building for Elephant Lands. Forest Hall contains behind-the-scenes elephant areas, a large open space where the animals can seek shelter from the elements, a visitor viewing gallery, a life-size tribute to Packy, and display cases detailing the five-thousand-year conjoined history of humans and elephants. Even Thonglaw's tusk—the one Roger hid amid the hay bales—was on display.

"This is the kind of shit I dreamed about when I was standing in your shoes," he told Lee. "The fact that you've actually got it now and can use it makes me jealous as hell." He turned in a slow circle. "The progress being made here is mind-boggling."

Lee thanked him, but added the changes were still less than what the elephant keepers hoped for. The facility was heavily value engineered, and many of the features they wanted got

Roger smiles as he sits at his old desk. (Photo by Melissa Crandall, 2018)

cut. Even so, they all understood the reality of life: the money well is not bottomless.

Lee escorted them deep into the back areas of the barn to see the isolation areas for sick animals, the permanent hoist should an elephant ever require lifting, and the new ERD. "Hey, Roger," he said as they strolled the upper level. "There's something here I think you should see."

Against a wall sat the battered wooden desk from the office in the old elephant barn, the spot where Roger had spent the majority of his waking hours writing reports, telling stories, conducting bull sessions, laughing, throwing a temper tantrum or two, smoking countless cigarettes, and drinking far too many cups of coffee.

"We couldn't let that get destroyed," Lee said. "It's a part of history."

Roger pulled out the chair and sat down. In an instant, the years rolled back and the ghostly outline of the old barn rose around him. Faintly, through the distance of years, he heard the caustic banter of keepers, the sound of elephants, the hum of hydraulic doors, and the knock-knock of Belle's trunk as she tried to lure him away from his work. He leaned back with an expression of contentment, savoring his memories of a time when a man in a dirty uniform and battered campaign hat walked with elephants.

He looked right at home.

A Concise Look at Roger's Elephants

Tracking captive elephants over a span of years can prove difficult because when they change hands, they're often given new names. Internet sites and the vast network of people devoted to elephants provided invaluable assistance in tracking what threads remain. Even so, some of the elephants Roger knew and loved have disappeared into the mists of time.

ROSY *Born 1947/49, Cambodia* *Died 1993, Portland*	Once sprained her ankle by trying to balance on her drinking trough.
TUY HOA *Born 1955, Vietnam* *Died 1983, Portland*	Appeared in the 11/24/61 issue of *LIFE* Magazine in a photo titled "The Nose Has It," in which she's kneeling at the edge of the dry moat, leaning almost to the tip-over point, to retrieve a slice of bread thrown by a child.
BELLE *Born 1952, Bangkok* *Died 1997, Portland*	The zoo's uber-babysitter, famous for being the mother of Packy.

THONGLAW *Born 1947, Bangkok* *Died 1974, Portland*	America's first breeding bull.
PET *Born 1955, Bangkok* *Died 2006, Portland*	When she died, zoo visitors honored her memory by making and decorating paper lanterns, which were hung in the trees surrounding the elephant barn. She was the first of the Oregon Zoo elephants to be buried in the elephant graveyard. At the time of her death, Pet was matriarch.
ME-TU *Born 1962, Portland* *Died 1996, Portland*	Christened "Thorny" and "Posy" before receiving her official name. There were thousands of entries, but Helen Thorsen of Delake, Oregon, won with her submission, a play on "Me, too!"
PACKY *Born 1962, Portland* *Died 2017, Portland*	Legend has it that upon his birth, the entire herd trumpeted a greeting. In 2013, a routine test for tuberculosis came up positive. He was treated with varying degrees of success, but proved resistant to medication. He is buried near Pet.
HANAKO *Born 1963, Portland*	Once jockeyed with Me-Tu for supremacy and broke off a tush inside the sulcus. Thirty minutes later, she did the same with the other one. Keepers trained her to raise her trunk so they could hose out the breaks and treat them with peroxide. Hanako developed a cancerous lesion on her left foot in 2018. Due to its size and location, surgery is not possible. Keepers have teamed with a veterinary oncologist and others to develop a long-term treatment plan. So far, results are encouraging. Hanako continues to reside at Point Defiance Zoo & Aquarium with her barn-mate, Suki.

EFFIE (aka SUE) *Born 1950, Thailand* *Died 1985, Florida*	As a calf, she lived with Sidney Snow, director of the Oakland Zoo, and enjoyed pleasure rides in the back of his car with her trunk hanging out the driver's-side window. As an adult, she roughed up a couple of handlers and snatched a woman's purse.
RAJAH (aka TEAK) *Born 1966, Portland* *Died 1978, Canada*	Roger remembers him as "cute, a roly-poly calf that looked just like his mother, Pet."
WINKIE *Born 1945* *Death unknown*	Imported to the U.S. as a five-year-old in 1950 by exotic animal dealer Henry Trefflich. Described as fat, contented, and mild-mannered, she was named for Henry Vilas Zoo director Fred Winkelman. It's reported she hurled mud at a photographer, and once ripped the door off her summer home and threw pieces of it at a night security guard.
HUGO *Born 1960, in the wild* *Died 2003, Portland*	While with Ringling, he participated in a Cherry Blossom Festival parade in Washington, DC, where he waded into the Lincoln Memorial reflecting pool, took a dip, and left behind an enormous bowel movement. Aggressive treatment for a systemic infection in 2003 brought no relief and he was euthanized.
DROOPY *Born 1968, Portland* *Died 1968, Portland*	Has the distinction of being the first elephant Roger ever touched.
DINO *Born 1963, Portland* *Died 1977, California*	Described as having a lot of spunk, but no coordination.

TINA *Born 1970, Portland* *Died 2004, Tennessee*	Tina spent fourteen years alone at the Vancouver Game Farm. In 1986, she gained a companion at last: Tumpe, a young female African elephant. They remained together until 2002, when Tumpe was sent to a zoo in the United States. Tina suffered from severe pododermatitis and degenerative osteoarthritis. She spent the last year of her life at The Elephant Sanctuary in Hohenwald, Tennessee. After she died, the herd stood vigil at her gravesite for two days.
JUDY *Born 1970, Portland*	After being sold to Wildlife Safari in 1972, she transferred to another owner in 1978. No further details are known.
GABRIEL *Born 1972, Portland* *Died 1983*	A press release early in his life placed him at Sterling Forest Gardens in Tuxedo, NY, performing in the "Wild Animal Circus Show." No further details are known.
STRETCH (aka TUSKANINI) *Born 1973, Portland* *Died 1976, Florida*	After his sale to an unspecified owner in Florida, no further details are known.
LOOK CHAI (aka SABU) *Born 1982, Portland* *Died 2012, California*	In 2010, Feld Entertainment (parent company of Ringling Bros.) donated Look Chai to the Performing Animals Welfare Society (PAWS) ARK 2000 elephant sanctuary in San Andreas, CA. He succumbed to the ravages of severe arthritis in multiple joints.
TUNGA *Born 1966, Thailand* *Died 1998*	No further information.

STONEY *Born 1973, Portland* *Died 1995, Nevada*	After being sold to Ken Chisholm in Canada, he was sold to animal trainers Mike and Sally LaTorres in 1975. He lived and performed with them until their divorce in 1986, at which point Mike LaTorres retained sole custody. In 1994, while practicing a hind-leg stand backstage at the Luxor Casino in Las Vegas, Stoney's left rear hamstring snapped. The injury was treated (there is heated debate as to how well) and he was placed in a sling to take the weight off the leg while it healed. When that proved unviable due to chafing from the straps that held him suspended, a special ERD was built to hold him. He remained in the ERD for almost a year, undergoing a fluctuating regimen of treatments, but his health deteriorated and he died. No necropsy was performed.
M&M (aka EMMA) *Born 1973, Portland* *Died 1986, Florida*	Sold in 1974, then went to Busch Gardens in 1975. She may later have been sold to a private owner. No other details are known.
TAMBA *Born 1969/70, Thailand* *Died 1993, Portland*	The shy elephant adored Roger to an embarrassing degree.
SUSI *Born 1952, Sri Lanka* *Died 1988, Portland*	Despite numerous breeding attempts, Susi failed to conceive. Her lack of interest in Tunga may have been a result of his biting off the end of her tail while in musth.

THONGTRII *Born 1979, Portland* *Died 1993 or 2003 (sources differ)*	No further information.
SUMEK (aka MY-OW) *Born 1978, Portland* *Died 1978, Portland (six weeks old)*	One of Hanako's short-lived calves.
KHUN CHORN (aka MR. ED) *Born 1978, Portland* *Died 2016, Missouri*	Renamed by staff at Dickerson Park Zoo, he lived there for thirty-six years. Zoo keepers taught him to throw a softball and he traveled to Meador Park to throw a ceremonial first pitch. Although given opportunities to breed, he never sired offspring. In 2016, staff found him lying in the barn. After using a variety of methods to encourage him to stand (including a hoist and rigging tractor tires to provide him with leverage) the fire department was called to assist with inflatable air bladders. All attempts were unsuccessful. He was euthanized the next day.
RAMA *Born 1983, Portland* *Died 2015, Portland*	Later in life, he developed a reputation as an abstract artist, creating designs with nontoxic tempera paints as a form of enrichment activity that were compared to Jackson Pollack, Joan Miró, and Paul Klee. His art has been exhibited at the Mark Woolley Gallery in Portland's Pearl District. In 2013, he tested positive for tuberculosis. He responded well to treatment, but the long-term effects of an old leg injury caught up with him. He is buried near Pet.

CHANG DEE (aka PRINCE) *Born 1987, Portland*	After many years of performing with Ringling, he retired to a farm run by the circus's parent company, Feld Entertainment. In 2011, he was donated to the PAWS ARK 2000 elephant sanctuary in California where he continues to reside.
ROSE-TU *Born 1994, Portland*	Rose-Tu continues to reside at Oregon Zoo, where she has delivered two calves: Samudra and Lily. Sadly, Lily died of EEHV in 2018.
CINDY *Date of birth unknown* *Died 2002, Washington*	After she returned to Tacoma, she made a three-cow herd with Suki and Hanako. Over time, crippling arthritis took its toll. When it became obvious she could no longer stand on her own, she was euthanized. The City of Tacoma mourned.
SUNG SURIN (aka SHINE) *Born 1982, Portland*	Grew into a big elephant like her father, Packy, and with her mother Pet's tendency to tease. During routine testing in 2017, she was diagnosed with tuberculosis. Quarantined, she successfully completed a lengthy round of treatment. Although never showing the slightest inclination to breed, she is an excellent midwife and auntie. She is the oldest and tallest member of the herd at the Oregon Zoo, and the matriarch.

Acknowledgments

Roger thanks his family for their love and support, and offers a tip of the hat to those friends who saved his bacon more than once: Joe Cochran, Mitch Finnegan, Murray E. Fowler, Larry Galuppo, Jay Haight, John Houck, Warren Iliff, Anna Keele, Meg Koonce, Gene Leo, Ann Littlewood, Fred Marion, Jack Marks, Steve and Jonolyn McCusker, Jill Mellen, Luke Metcalf, Gordon Noyes, John Pasco, Wes Peterson, Bets Rasmussen, McKay Rich, Denslow Robbins, Charlie Rutkowski, Jim Sanford, Anne Moody Schmidt, Bill Scott, Roland Smith, Sioux Strong, Dave Thomas, and Al Tucker. His apologies to anyone he's inadvertently overlooked.

Roger Henneous braved the demons of his past to partner with me in telling his story. I love and admire the man, and am proud to call him my friend.

Thank you to RoseMerrie Henneous, who welcomed me into their home and encouraged us every step of the way. Thank you to their daughters, Michelle Plaschka and Melissa Mayes, and to Donald and Beverly Henneous.

My gratitude to everyone at Ooligan Press for championing this story. Special thanks to Alyssa Schaffer and Joanna Szabo, Hope Levy and Amylia Ryan, Brennah Hale and Zoe LaHaie, Ari Mathae and Taylor Thompson, Abbey Gaterud, Monique Vieu, Madison Schultz, Julie Collins, Jenny Kimura and Laura

Mills, and Melinda Crouchley. My sincere apologies to anyone I've forgotten.

Thanks to those who shared their knowledge, expertise, and enthusiasm:

At the Oregon Zoo: Bob Lee, Mitch Finnegan, Charlie Rutkowski, Diana Bratton, Margot Monti, Rick Haynes, Ivan Ratcliff, Jim Rorman, Shawn Finnell, Gilbert Gomez, Pam Starkey, Carlos R. Sanchez, DVM, MSc, Matt Miles, Tarah Bedrossian, and Joseph Sebastiani.

At the Point Defiance Zoo and Aquarium: John Houck, Shannon Smith, and Kate Burrone.

At the Dickerson Park Zoo: Michael Crocker and Melinda Arnold.

Special thanks to: Betty Bernt at the Oregon Department of Corrections; Sergeant Brown of the Multnomah County Sheriff's Work Crew Unit; John Carroll and Michael Gross at The Authors Guild; Larry Galuppo, DVM, DACVS, UC Davis School of Veterinary Medicine; Kim Gardner at the Performing Animals Welfare Society (PAWS); Jay Haight; Mary Hansen at the City of Portland Archive and Records Center; Therese Kopiwoda; Brittany Lester; Todd Montgomery at The Elephant Sanctuary in Tennessee; Linda Reifschneider at Asian Elephant Support; Sioux Strong; Ashlyn Templeton at the San Antonio Zoo; and April Yoder at the Elephant Managers Association.

Thanks as well to: Rolanda Kaiser Andrade at Circus Vargas; Dan Brands at Wildlife Safari; Greg Geiser, Esquire; Chris Havern, staff historian for the United States Coast Guard; Daryl Hoffman at the Houston Zoo; Ashley Mayrianne Jones at the Association of Zoos and Aquariums; Zachary Kretchmer, Esquire; Ann Littlewood; Linda Lutes at *The Columbian*; Dennis McCormick at *Capital Newspapers*; Jennifer McFadden and Elaine Graves at *National Geographic*;

David Muir and Adrienne Southard of Information and Referral for the City of Portland; Beth Nakamura, Drew Vattiat, and Michael Zacchino at *The Oregonian*; Coco Noelle at Danner Boots; Ryan Petit at the Bureau of Labor Statistics; and Laura Schillinger.

My deep gratitude to John Valeri for the Sanity Sessions.

Thanks to Ed Everett, Dan Foley, Terry George, Pam Hohmann, John Houck, Stacey Longo, Jim Mastro, Linda Reifschneider, Risë Shamansky, and John Valeri, who read various versions of the manuscript and offered invaluable insight.

Ed Everett deserves extra kudos for author support. He must wonder sometimes what the hell he got into by marrying a writer, but I'm delighted to report that after more than four years of listening to me talk about little except elephants, he's been totally corrupted and can now only think of one animal in particular when he hears the phrase, "Me, too."

References

"2nd Elephant Baby Born In Portland." *The Oregon Journal*. October 3, 1962.

"3rd Pachytot Named Dino." *The Oregon Journal*. October 5, 1963.

Alexander, Shana. "For the Love of Elephants." *LIFE Magazine*. March 1980.

Alexander, Shana. *The Astonishing Elephant*. Random House. 2000.

Alexander, Shana. "Belle's Baby — 225 Pounds and All Elephant." *LIFE Magazine*. May 11, 1962.

Alexander, Shana. "I go back to an elephant nursery." *LIFE Magazine*. April 7, 1967.

"Animal Enrichment." Smithsonian Conservation Biology Institute. www.nationalzoo.si.edu. Retrieved 2015.

"Apple-Ball Helps Zoo Elephants Stay Fit." *Los Angeles Times*. May 10, 1987.

Arave, Lynn. "Hogle Zoo history full of cute, cuddly, cantankerous critters." *Deseret News*. September 19, 2010.

"Asian Elephant." World Wildlife Fund. www.worldwildlife.org

"Asian Elephant and African Elephant: Endangered Species." Bagheera. 2016. www.bagheera.com

"Asian Elephant Conservation Fund." U.S. Fish & Wildlife Service, International Affairs. 2015. www.fws.gov

Asian Elephant Specialist Group Newsletter, Number 7. International Union for Conservation of Nature and Natural Resources. Autumn 1991.

"Asian elephants." WWF Global. 2016. www.panda.org

"Asian Elephants: Threats and Solutions." American Museum of Natural History. 2016. www.amnh.org

"Asian elephants at the Oregon Zoo in Portland." 2001. www.asianelephant.net

"Auction planned for elephant tusk." *Prince George Citizen*. May 10, 1978.

"Baby elephant dies." *Eugene Register-Guard*. May 1, 1978.

Baskas, Harriet. *Oregon Curiosities: Quirky Characters, Roadside Oddities, and Other Offbeat Stuff*. Globe Pequot Press. 2003.

"Bear Mauls Zoo Worker." *Spokane Daily Chronicle*. September 1, 1966.

Beebe, Craig. "Changes complicate Packy's TB treatment, shed light on elephant aging." *Metro News*. December 8, 2014.

Berens, Michael J. "Elephant havens face zoo-industry backlash." *The Seattle Times*. December 2, 2012.

Berens, Michael J. "Elephants are dying out in America's zoos." *The Seattle Times.* December 1, 2012.
Berger, Joe. "Third Elephant Born at Portland Zoo." *Portland Journal.* Undated clipping from collection of Roger Henneous.
Bernstein, Adam. "Philip W. Ogilvie, 70." *The Washington Post.* September 6, 2002.
"Bio: Roland Smith." www.rolandsmith.com
Blakemore, Erin. "Lincoln Turned Down a Chance to Fill the U.S. With Elephants." www.mentalfloss.com. January 13, 2017.
"Body of trainer recovered from guard of bull elephant." *Eugene Register-Guard.* June 28, 1979.
"Boy Who Saw Killing Says Elephant 'Scared.'" *The Evening Independent.* June 29, 1966.
Bradley, David. "Elephant Sex." *Reactive Reports.* 2009. www.reactivereports.com
"Bull elephant ends period of isolation." *Eugene Register-Guard.* August 20, 1969.
Butler, Jess. Letter to Morgan Berry. Undated. From the collection of Roger Henneous.
Castano, Carla. "Elephant bull hook: insurance, weapon or both?" KOIN 6. May 5, 2015.
Castano, Carla. "Why Oregon wants to breed elephants." KOIN 6. May 3, 2015.
Chambers, Dave. "Chemical castration of bull elephants fast becoming an option." *Business Day.* September 19, 2017. www.businesslive.co.za.
"Cindy explores new San Diego home." *The Spokesman-Review.* December 16, 1982.
Colby, Richard. "Death of Thonglaw blamed on drug reaction." *The Oregonian.* No date given. From the collection of Roger Henneous.
Corp-Minamiji, DVM. "Dr. Murray Fowler's legacy lives in those he inspired." VIN News Service. May 29, 2014.
Cowlitz County Law Enforcement History Project. "The Elephant, the Photographer, and the Sheriff." October 23, 2009. Cclehistory.blogspot.com
Croke, Vicki. *The Modern Ark: The story of Zoos: Past, Present and Future.* Scribner. 1997.
Croke, Vicki. "When This Baby Elephant Cried, Mom and Super-Auntie Rescued." October 7, 2014. www.thewildlife.wbur.org
Croke, Vicki Constantine. *Elephant Company.* Random House. 2014.
"DPZ Euthanizes Bull Elephant 'Mr. Ed.'" OzarksFirst.com. January 19, 2016.
"'Dangerous' Elephant To Get Bigger Quarters." *The New York Times.* November 26, 1982.
"Defiant attitude better suited for Point Defiance." *The Seattle Times.* August 12, 2005.
"Delake Woman's Entry Selected From Thousands For Elephant." *The Oregon Journal.* November 2, 1962.
Denyer, Simon. "China to ban ivory trade within a few years as pressure mounts in Hong Kong." *The Washington Post.* October 21, 2015.
"'Difficult' elephant moving to Tacoma." *Eugene Register-Guard.* December 4, 1997.
"Doctor hoped boot would help elephant." *The Oregonian.* Date unknown. From the collection of Roger Henneous.
Doty, Anna. "Extinction Breeds Innovation: A History of the Oregon Zoo." Undergraduate Thesis, Concordia University. 2017.

Dresbeck, Rachel. *Insider's Guide® to Portland, Seventh Edition*. Insider's Guide. 2011.

Durrow, Heidi. "Zoo's Tamba Takes A Tumble." *The Oregonian*. September 20, 1990.

"Elephant and Bees Project." Save the Elephants. www.elephantsandbees.com

"Elephant Breaks Trainer's Arm." *Eugene Register-Guard*. July 5, 1969.

"Elephant, Calf Not Wanted." *Ellensburg Daily Record*. October 3, 1968.

"Elephant Man's Death Puzzle." *Spokane Daily Chronicle*. June 28, 1979.

"Elephant matriarch Rosy dies at 43." *The Oregonian*. January 29, 1993.

"Elephant May Prefer To Breed In Private." *The Seattle Times*. July 28, 1991.

"Elephant Reproduction." Association of Zoos and Aquariums. June 2009. www.elephanttag.org

"Elephant Touch." *Desert Sun*. July 20, 1973.

"Elephant Visits Portland." *Spokane Daily Chronicle*. April 12, 1967.

"Elephants at the Oregon Zoo: A History." Oregon Zoo. www.oregonzoo.org

"Elephants — Longevity and Causes of Death." SeaWorld Parks & Entertainment. 2016. www.seaworld.org

"Elephus maximus." International Elephant Foundation. www.elephantconservation.org

"Enrichment." Association of Zoos and Aquariums. www.aza.org

Evans, Frank Kinsey. *You're the Elephant Man: If You're Still Alive After Two Weeks, You're Permanent*. Writer's Showcase, iUniverse.com. 2000. Pp. 99, 104.

Fabricius, Karl. "When Geese Attack!" May 27, 2010. Scribal.com.

"Famed Elephant Dead." *Circus Report*. December 9, 1974.

Federman, Stan. "Packy comes of age for grand birthday party." *The Oregonian*. April 7, 1983.

Federman, Stan. "Macho pachyderm Hugo 'thinks he's died, gone to heaven' at zoo." *The Oregonian*. June 3, 1984.

Finnegan, Mitch and Margot Monti. "Surgical Management of Phalangeal Osteomyelitus in a Female Asian Elephant (Elephus Maximus)." *The Elephant's Foot: Prevention and Care of Foot Conditions in Captive Asian Elephants*. John Wiley & Sons. April 2008.

Foerner, Joseph J., Richard I. Houck; John F. Copeland; Michael J. Schmidt; H.T. Byron; and John H. Olsen. "Surgical Castration of the Elephant (*Elephas Maximus* and *Loxodonta Africana*)." *Journal of Zoo and Wildlife Medicine*. Vol. 25, No. 3, September 1994.

"Four Of A Kind Makes Full House At Portland's City Zoo." *The Oregonian*. September 25, 1963.

Fowler, Murray E. "Castration of an Elephant." *The Journal of Zoo Animal Medicine*. Vol. 4, No. 3, September 1973.

Fowler, Murray E. "An Overview of Foot Condition in Asian and African Elephants." First North American Conference on Elephant Foot Care and Pathology. 1998.

Fowler, Murray E., Eugene P. Steffey; Larry Galuppo; and John R. Pascoe. "Facilitation of Asian Elephant (Elephas maximus) Standing Immobilization and Anesthesia with a Sling." *Journal of Zoo and Wildlife Medicine*. Volume 31, No. 1, pp 118–123. March 2000.

Frazier, Amy, and KOIN 6 News Staff. "After years, Oregon Zoo's Elephant Lands opens." December 15, 2015. www.koin.com.

Frazier, Laura. "250 pounds of elephant ivory from Oregon Zoo collection pulverized in Times Square." *The Oregonian*. June 19, 2015.
Fritsch, Jane. "Beatings, Abuse: Elephants in Captivity: a Dark Side." *Los Angeles Times*. October 5, 1988.
Fritsch, Jane. "Elephant handling faces examination." *The Oregonian*. October 17, 1988.
Galuppo, Larry, DVM. Telephone interview and email correspondence with author. 2015, 2016.
"Getting to know Packy." Oregon Zoo. www.oregonzoo.org
Gipson, Chester A. Letter to 'Concerned Citizen.' United States Department of Agriculture. November 2004. www.elephants.com
Glenn, Stacia. "Elephant diagnosed with cancer at Point Defiance Zoo & Aquarium." *The News Tribune*. March 22, 2018.
"Goose attack leaves Ottawa cyclist shaken and scarred." *CBC News*. June 25, 2014. www.cbc.ca
"Greeting." *Oregon Journal*. February 22, 1966.
Haight, Jay. "Captive Management of Breeding Asian Elephants." *Asian Elephant Specialist Group Newsletter*. Autumn 1991.
Haight, Jay. "Healthy animals are happy, at times, doing nothing." October 15, 2000. www.oregonlive.com
Haight, Jay, Roger Henneous, and Douglas Groves. "Specialized Tools for Elephant Foot Care." *Recent Developments in Research and Husbandry at the Washington Park Zoo*. Jill Mellen and Ann Littlewood, Editors. 1981.
Harden, Kevin. "Packy laid to rest in elephant graveyard." *Portland Tribune*. April 13, 2017.
Harvey, John. "Surprise Arrival Of Third Baby Elephant Enlivens Busy Zoo." *The Oregonian*. September 17, 1963.
Harvey, John. "Zoo Welcomes Baby Elephant." *The Oregonian*. September 16, 1963.
"Has trunk, won't travel: Lily belongs to Oregon Zoo." February 7, 2013. www.oregonzoo.com
Henneous, Roger L. and Diana Barker. "Portland: Elephant Town USA." *The Portland Magazine*. March 1981.
Henneous, Roger. Letter to Gary L. Dunn, Tulsa Zoological Park. May 20, 1976.
Henneous, Roger. Thank you letter to citizens of Metropolitan area. March 5, 1998.
Henneous, Roger, Michael J. Schmidt, and Jay D. Haight. "Deadly Dilemmas of Captive Elephant Breeding." From the collection of Roger Henneous.
Hermes, R.;Saragusty, J.; Schaftenaar, W.; Göritz, F.; Schmitt, DL; and Hildebrandt, TB. "Obstetrics in elephants." *Theriogenology*. July 15, 2008.
Hernandez, Tony. "Oregon Zoo's Rama, a 31-year-old elephant, put to sleep after decades-long injury." March 30, 2015. www.oregonlive.com
Hill, Richard L. "Belle's Gone — but not forgotten." *The Oregonian*. Undated. From the collection of Roger Henneous.
Hill, Richard L. "Public joins in mourning death of popular Belle." *The Oregonian*. April 24, 1997.
Hillinger, Charles. "With 1,000 Enrolled, This Kids' Camp Is a Real Zoo." *Los Angeles Times*. July 24, 1989.
"History." Oregon Zoo. www.oregonzoo.org

Holt, Nathalia. "The Infected Elephant in the Room." *Wild Things*, Slate's Animal Blog. March 24, 2015.

"Human Elephant Conflict." EleAid. 2016. www.eleaid.com

Hune-Brown, Nicholas. "What the Elephants Know." The Elephant Sanctuary in Tennessee. www.elephants.com

Hunter, Mary. "An Elephant Never Forgets (Ever?)." www.stalecheerios.com

Huntington, Rebecca. "How do elephants spell relief?" *Eugene Register-Guard*. December 28, 1998.

"Inmate Work Crew Information." DOC Operations Division: Prison. www.oregon.gov. Retrieved 2015.

"In Memory of Frederic W. Marion." Pable-Evertz Funeral Home. www.pableevertzfuneralhome.com

International Elephant Foundation. www.elephantconservation.org/

"Isolated Cindy Prompts Debate Over Elephant Care." *Seattle Times*. December 27, 1992.

"Issues: Human-elephant conflict. India: up to 300 people may be killed annually in human-elephant conflicts." WWF Global. www.panda.org

Jaynes, M. *Elephants Among Us: Two Performing Elephants in 20th Century America*. Earth Books. 2013. Pp. 1–114, 161–164

Kalk, Penny and Chris Wilgenkamp. "Elephant Foot Care Under the Voluntary Contact System: Problems and Solutions." First North American Conference on Elephant Foot Care and Pathology. March 1998.

Keene, Linda. "Handlers Face Elephantine Risks — Statistically, Pachyderm Training One Of Most Dangerous Jobs." *The Seattle Times*. July 26, 1994.

Kern, David. "Elephant Man." *The Columbian*. April 19, 1978.

King, Barbara J. *How Animals Grieve*. University of Chicago Press. April 2014. Pp 60–63.

Kuralt, Charles. "On the Road, Episode 12." Circa 1983.

Latham, Brian. "The Number of Elephants in Parts of Zimbabwe Is Plunging." *Bloomberg*. February 18, 2015.

Laquedem, Isaac. "Warren Iliff (1936–2006), the man who loved animals." www.isaac.blogs.com

LePage, Andrew. "Tacoma to Take Back Cindy, an Albatross of an Elephant for Park." *Los Angeles Times*. March 2, 1989.

"Lincoln Rejects the King of Siam's Offer of Elephants." www.battlefield.org

"Lincoln to Thai King: Thanks but no thanks for the elephants." www.nydailynews.com. March 24, 2018.

"Lioness kills teenager in Portland zoo mauling." *The Bulletin*. July 6, 1970.

"Longtime zoo chief dies at 75." *Eugene Register-Guard*. August 10, 1987.

"Look, Mom, It's More Elephants." *LIFE*. November 2, 1962.

Lyon, Ann. Letter to *The Oregonian*. October 17, 1987.

Mailman, Erika. *Images of America: Oakland Hills*. Arcadia Publishing. November 2004. Pp 120.

Maluy, Patrick D. Interview with Roger Henneous. Woodland Park Zoo. February 1998.

"Man sentenced for killing lions in Portland Zoo." *The Oregonian*. February 7, 1973.

Mansfield, Duncan. "Pachyderm Discovers New Life Through Painting." *Los Angeles Times*. September 7, 1997.
Mar, Khyne U., Mirkka Lahdenperä, and Virpi Lummaa. "Causes and Correlates of Calf Mortality in Captive Asian Elephants (Elephas maximus)." *PLOS one*. March 1, 2012. https://journals.plos.org
Markowitz, Hal, Michael Schmidt, Leonie Nadal, and Leslie Squier. "Do Elephants Ever Forget?" *Journal of Applied Behavior Analysis*. 1975, 8; 333–335. Number 3. Fall 1975.
Marr, John. "The Saga of a Man, Two Lions, and a Freak Accident That Got Freakier." Gizmodo. July 22, 2015. www.gizmodo.com
Mascarenhas, Hyacinth. "13 species we might have to say goodbye to in 2015." *Global Post*. January 1, 2015.
Mattila, Walt. "Our Belle's Bouncing Baby Boy Ends Portland Zoo's Suspense." *Portland Reporter*. April 14, 1962.
Mattila, Walt. "Packy Becomes Unruly; Elephant 'School' Hinted." *Portland Reporter*. February 26, 1963.
Maves, Jr., Norm. "Last days with the elephants." *The Oregonian*. February 24, 1998.
McClendon, Jessie. "Packy the elephant (1962–)" The Oregon Encyclopedia. www.oregonencyclopedia.org
McLellan, Dennis. "Warren Iliff, 69; Aquarium of the Pacific's First Chief Executive." *Los Angeles Times*. August 14, 2006.
"Meet the Elephants." Performing Animals Welfare Society. www.pawsweb.org
Mikota, Susan K., DVM. "A Brief History of TB in Elephants." Elephant Care International. Undated. www.elephantcare.org.
Miller, Robert H. DVM. "Mind Over Miller: An elephant-size job." *Veterinary Medicine*. April 1, 2006. www.veterinarymedicine.dvm360.com.
Monroe, Linda Roach. "A hint for zoo keepers: Never trust a bull elephant." *The Oregonian*. Undated clipping from collection of Roger Henneous.
Moore, Sandra D. "Zoo keeper of year: He likes elephants better than people." *The Community Press*. June 5, 1974.
"More orphaned elephants means more milk." The David Sheldrick Wildlife Trust. www.sheldrickwildlifetrust.org. October 18, 2013.
Moss, Ruth. "Today's zoo: If Noah only gnu!" *Chicago Tribune*. April 27, 1974.
"Much-loved elephant, 'Pet,' euthanized at zoo." Oregon Zoo. August 21, 2006. www.katu.com
Muldoon, Katy. "In the big business of elephants, breeding is a key issue, Oregon Zoo finds." *The Oregonian*. December 9, 2012.
Muldoon, Katy. "Oregon Zoo elephant's tuberculosis diagnosis a challenge for him and his caretakers." *The Oregonian*. June 6, 2013.
Muldoon, Katy. "Surgeons ready zoo's Belle for removal of infected toe." *The Oregonian*. March, 19, 1997.
"National Woman Killed by Elephant." *The Tuscaloosa News*. May 3, 1978.
O'Brien, Les. "Chemical Signals." LesOBrien.com. www.lesobrien.com
Ogilvie, P.W., Ph.D. "Animals of the Month — Portland's Elephants." *PZG Newsletter*. December 1974, Volume 3, Number 11.

O'Neill, Patrick. "Memorial service honors great gray lady." *The Oregonian*. No date given. From the collection of Roger Henneous.
"Ordinance authorizing the execution of an agreement with the Portland Zoological Society for management of the Portland Zoo." Calendar No. 2111. Ordinance No. 132821. City of Portland. June 1971.
"Oregon Zoo awarded for innovative elephant pool." KGW.com. July 12, 2016.
"Oregon zoo elephant Packy not reacting well to active tuberculosis treatment." *Daily News*. January 26, 2014.
"Oregon zoo elephant Packy struggling with TB." *New York Daily News*. January 26, 2014.
"PAWS saddened by loss of gentle giant Sabu, our oldest Asian bull elephant." PAWS Press Release. January 13, 2012.
"Pachyderm Discovers New Life Through Painting." *Los Angeles Times*. September 7, 1997.
"Pachyderm Pending. Oregon Zoo Elephant Birth Expected in Late 2012." Information and Resource Kit. www.oregonzoo.org. 2012.
"Pachyderm Talk in Portland." *Seattle Times*. September 15, 1995.
"Packy hurts zoo keeper." *The Bulletin*. October 11, 1977.
Painter, Jr., John. "Packy's Big Birthday Bash." *The Oregonian*. April 10, 1992.
Parks and Recreation, The City of Portland, Oregon. 1941–1960. www.portlandoregon.gov
Patel, Trishula. "For Elephants in Zimbabwe, a Deeply Troubling Present and Future." *PassBlue*. March 17, 2015.
Pement, Jack. "Prize Specimens Join Zoo Family." *Oregon Journal*. October 1, 1975.
"Pensive Pachydermic Pleas Attend Baby's Weigh-In." *The Oregonian*. September 17, 1963.
Performing Animals Welfare Society. www.pawsweb.org
Petersen, Carolyn. "Beer Kegs and Bowling Balls Help Amuse Zoos' Big Lions and Tigers and Bears." *Los Angeles Times*. October 25, 1987
Peterson, Lacey. "Sabu, the oldest elephant at PAWS, dies at 29." *The Union Democrat*. January 13, 2012.
"Point Defiance getting another unpredictable pachyderm." *Ellensburg Daily Record*. December 2, 1997.
"Point Defiance Zoo I: Hanako." Show Me Elephants. April 8, 2011. www.showmeelephants.com
"Point Defiance Zoo & Aquarium and Oregon's Metro Washington Park Zoo Announce Transfer of Asian Elephant." *PR Newswire*. December 1, 1997.
"Popular Elephant's Carcass Dumped With The Garbage." KOMOnews. August 31, 2006. www. komonews.com
"Portland boasts elephant baseball." *The Bulletin*. April 19, 1987.
"Portland may end up with pair of unwanted elephants." *The Bulletin*. October 2, 1968.
"Portland proud of Belle's baby elephant." *Ellensburg Daily Record*. April 19, 1962.
"Portland stuck with Effie." *Eugene Register-Guard*. November 13, 1968.
"Portland Zoo — Droopy." Show Me Elephants. May 4, 2011. www.showmeelephants.com
"Portland zoo director named." *Eugene Register-Guard*. January 9, 1988.

"Portland zoo gets curator from Chicago." *The Spokesman-Review*. February 19, 1987.
"Portland zoo has new baby elephant." *The Bulletin*. March 16, 1978.
"Portland zoo keeper critically hurt." *Eugene Register-Guard*. December 7, 1990.
"Portlanders sympathetic for recuperating elephant." *Eugene Register-Guard*. March 30, 1997.
"'Prince Utah,'" Baby Elephant of Salt Lake City." *Our Dumb Animals*, Volumes 51–53, page 37. Massachusetts Society for the Prevention of Cruelty to Animals. 1918.
"Proceedings of the First North American Conference on Elephant Foot Care and Pathology." Portland, Oregon. November 1998.
"Pushy elephant loses struggle." *The Victoria Advocate*. January 26, 1992.
"Ralph Mitchell Zoo, Independence, Kansas." Show Me Elephants. www.showmeelephants.com
"Rama (1983–2015)." The Rama Exhibition. www.theramaexhibition.com
"Rama the elephant is being treated for tuberculosis." Oregon Zoo. June 1, 2013. www.oregonzoo.org
"Rama the Oregon Zoo elephant dies." KOIN6. March 30, 2015. www.koin.com
Rasmussen, L.E.L., H.S. Riddle, and V. Krishnamurthy. "Mellifluous matures to malodorous in musth." *Nature*. 415, pp 975–976. February 28, 2002.
Rasmussen, L.E.L., and V. Krishnamurthy. "How Chemical Signals Integrate Asian Elephant Society: The Known and the Unknown." *Zoo Biology*. Volume 19, Issue 5, pp 405–423. 2000.
Rasmussen, Lois E., Michael J. Schmidt, Roger Henneous, Douglas Groves, and G. Doyle Daves, Jr. "Asian Bull Elephants: Flehmen-Like Responses to Extractable Components in Female Elephant Estrous Urine." *Science*. Volume 217, Issue 4555, pp 159–162. July 1982.
Ray, Nancy. "Witnesses Say Elephant Was Striking Back." *Los Angeles Times*. December 22, 2011.
Regan, John M. "The Point Defiance Zoo — Conversion to Protected Contact and Handling Problem Elephants." *AAT Magazine*. 2000.
Reynolds, Ed. "How to exercise a lazy elephant." *National Examiner*. June 16, 1987.
Richards, Leverett. "Our Rosy's A Mother; Wee Girl Elephant Born." *The Oregonian*. Date-stamped October 1962. From the collection of Roger Henneous.
Richards, Leverett. "'Rosy' Gives Birth To 2nd Daughter; Newcomer Tips Scales At Petite 210." *The Oregonian*. December 4, 1965.
Richards, Leverett. "Thonglaw's trainer mourns death of famous zoo elephant." *The Oregonian*. Undated clipping from the collection of Roger Henneous.
Richards, Leverett. "New Baby Crowds Nursery." *The Oregonian*. Undated clipping. From the collection of Roger Henneous.
Richards, Leverett. "Girl Latest Addition To Elephant Family." *The Oregonian*. October 4, 1962.
Richards, Leverett. "Elephant Young Like Those Of All Species — Mischievous." *The Oregonian*. July 8, 1962.
Riddle, Heidi S., Bets Rasmussen, and Dennis L. Schmitt. "Are captive elephants important to conservation?" *GAJAH, Journal of the Asian Elephant Specialist Group*. Number 22. July 2013. Pp 57–61.

"Ringling Bros. and Barnum & Bailey Donates Two Retired Asian Bull Elephants to the Performing Animal Welfare Society (PAWS)." Feld Entertainment. ww.feldentertainment.com

Rodriguez, Bonnie. "PAWS saddened by loss of gentle giant." *The Galt Herald Online*. January 19, 2012.

Roe, Amy. "Free the elephants! Problems plague the Oregon Zoo's pachyderms." *Willamette Week*. August 28, 2001.

Rollins, Michael. "Packy's birth in 1962 captivated the United States." *The Oregonian*. August 25, 2008.

"Rosie Joins Tailwaggers." No source. Date-stamped October 12, 1958. From the collection of Roger Henneous.

Ross, Winston. "Elephant Calf Lily Rescued by Oregon Zoo." *The Daily Beast*. February 12, 2013.

"Rosy Off Launching Pad With Portland's Second Pachyderm." *The Oregonian*. October 4, 1962.

Rutkowski, Charlie, Fred Marion, and Ray Hopper. "A Discussion of Three Types of Foot Problems: Split Nail, Abscesses, and Cuticular Fluid Pockets." First North American Conference on Elephant Foot Care and Pathology. 1998.

Saxe, John Godfrey. "The Blind Men and The Elephant." www.noogenesis.com

Schaul, Jordan Carlton. *Elephants in Captivity: A Perspective from Former AZA Director/William Conway Chair of Conservation and Science*. National Geographic. May 5, 2013.

Schmidt, Michael J., DVM. "Technic and Applications of Venipuncture in the Elephant. Recent Developments in Research and Husbandry at the Washington Park Zoo." 1981. Jill Mellen and Ann Littlewood, Editors.

Schwartz, Mark. "Elephants Pick up Good Vibrations — Through Their Feet." Stanford University News Service. March 5, 2001.

"Science is in: Zoo on right track with elephant lands." *Metro News*. July 14, 2016.

Scribner, Brad. "100,000 Elephants Killed by Poachers in Just Three Years, Landmark Analysis Finds." *National Geographic*. August 18, 2014.

"She Eavesdropped on Elephants." *Stanford Magazine*. January/February 2007.

Sheldrick, Daphne, D.B.E. "Elephant Emotion." www.sheldrickwildlifetrust.org

Sheldrick, Daphne, D.B.E. *Love, Life, and Elephants: An African Love Story*. Picador. 2010. Pp. 76.

Simnacher, Joe. "Warren J. Iliff: Visionary credited with saving Dallas Zoo." *The Dallas Morning News*. August 6, 2006.

Simmons, Sunshine. "Packy's Tuberculosis treatment proves to be a struggle." Examiner.com. January 28, 2014.

Skalicky, Michele. "Member of Dickerson Park Zoo's Elephant Herd Dies." KSMU.org. January 19, 2016.

Skinner, Nicole. "African elephant numbers collapsing." *Nature. International Weekly Journal of Science*. August 19, 2014.

Smith, Carol. "Elephant companions mourn passing of Tina." *Seattle Post-Intelligencer*. July 26, 2004.

Smith, Rachael. "Oregon Zoo." The Oregon Encyclopedia. www.oregonencyclopedia.org

Smith, Rod. "Elephant Capitol for the World." *News-Tribune*. Undated. From the collection of Roger Henneous.

Smith-Spark, Laura. "Why do elephants have hair on their heads? Scientists solve head-scratcher." CNN. October 18, 2012.

Sorel, Lawrence. "Thoughts on how we use language." LinkedIn. April 21, 2016.

Sterling Forest News Service. Press release, undated (circa April 1974). From the collection of Roger Henneous.

Stewart, Ed. "No Ethical Way to Keep Elephants in Captivity." *National Geographic*. May 3, 2013.

"Strange farm animal." *The Southeast Missourian*. August 12, 1982.

"Studbook Mysteries — Lincoln Park Zoo." Show Me Elephants. April 12, 2011. www.showmeelephants.com

Sundell, Alison. "Cancer diagnosis for Tacoma elephant hits zoo staff hard." K5News. March 22, 2018.

"Suzy (Susie) at Portland Zoo (Metro Washington Park Zoo)." www.elephant.se

Teagle, Andrea. "Zimbabwe plans to sell baby elephants to global buyers." *The Guardian*. January 15, 2015.

"Temperamental elephant to return to Tacoma." *Eugene Register-Guard*. February 18, 1992.

"Thailanders Visit Rosy: Stylish Home Amazes Visitors." Source unknown. Date-stamped September 6, 1957. From the collection of Roger Henneous.

"The Nose Has It." *LIFE Magazine*. November 24, 1961. Pp. 198

Thompson, Jeff and KGW.com Staff. "Oregon Zoo elephant 'Rama' euthanized." KGW.com. March 31, 2015.

"Tina." The Elephant Sanctuary in Tennessee. www.elephants.com

Tisdale, Sallie. "The Only Harmless Great Thing." *The New Yorker*. January 23, 1989.

Tisdale, Sallie. "The Birth." *Portland Magazine*. Spring 2003.

"Tom Packs Elephants." Show Me Elephants. www.showmeelephants.com

"Tots name baby elephant 'Dino.'" *The Spokesman-Review*. October 6, 1963.

"Tow Trucks Give Elephant A Lift." *The Seattle Times*. September 20, 1990.

"Tunga debuts as lead elephant at Portland Zoo." *Eugene Register-Guard*. December 6, 1979.

"Tuy Hoa Rewrites Zoological Books, Sets Gestation Record." *The Oregonian*. February 22, 1963.

Vecchio, Tony. "Zoo cares deeply and works hard to stimulate animals." October 15, 2000. www.oregonlive.com

Watson, Lyall. *Elephantoms: Tracking the Elephant*. W.W. Norton & Company. 2002.

Watt, Teresa. "Asian elephant at Point Defiance Zoo diagnosed with cancer." KOMONews.com. March 22, 2018.

"We Have Contact." The Zoo Review: Insights into the World of Zoos and Aquariums. August 30, 2015. thezooreviewer.blogspot.com/

"Welcome Prince! Prince joins Nicholas and Sabu on Bull Mountain." PAWS E-News, July 26, 2011.

Wentz-Graff, Kristyna. "Oregon Zoo's 'Elephant Lands' habitat now complete." *The Oregonian*. December 15, 2015.

Wheeler, Linda and Scott Bowles. "First Elephant Born At National Zoo Dies." *The Washington Post*. April 27, 1995.

Whitfield, John. "Honey smells like teen elephant." *Nature*. February 28, 2002.
"Why Care?" World Elephant Day. 2016. www.worldelephantday.org
Wilkie, Gavin S., Andrew J. Davison, Karen Kerr, et al. "First Fatality Associated with elephant Endotheliotropic Herpesvirus 5 in an Asian Elephant: Pathological Findings and Complete Viral Genome Sequence." *Scientific Reports*, 4. Article Number 6299 (2014). September 9, 2014.
Willard, Sherri J. "Preserving the Asian elephant." *The Oregonian*. July 26, 1990.
Williams, Kate. "Packy, the much-loved Oregon Zoo elephant, dies at 54." *The Oregonian*. February 9, 2017.
Wollheim, Bob. "Roger Henneous: He's the world's luckiest as his brother's elephant keeper." *Metropolis*. January 1978. Volume 6, Number 4.
Wood, Mark Dundas. "Packy comes of age." *Willamette Week*. April 5–11, 1983.
Yoder, April. Email correspondence with author. 2016.
Zachariah, Arun; Jian-Chao Zong, Simon Y. Long, et al. "Fatal Herpesvirus Hemorrhagic Disease in Wild and Orphan Asian Elephants in Southern India." *Journal of Wildlife Disease*. 2013; 49(2): 381–393. doi: 10.7589/2012-07-193.
"Zoo Attendant Recovering After Mauling by Bear." *Eugene Register-Guard*. September 2, 1966.
"Zoo curator resigns to take Arizona job." *Eugene Register-Guard*. October 27, 1986.
"Zoo elephant Winkie grabs, kills girl, 3." *Wisconsin State Journal*. June 29, 1966.
"Zoo keeper hurt." *The Oregonian*. February 18, 1977.
"Zoo mourning death of elephant." Dickerson Park Zoo. 2016. www.dickersonparkzoo.org
"Zoo Officials Spare Killer Elephant." *The Capital Times*. June 19, 1966.
"Zoo says goodbye to beloved Asian elephant Rama." Oregon Zoo. March 30, 2015. www.oregonzoo.org
"Zoo wins award for elephants." *The Oregonian*. October 12, 1976.
"Zoo's elephant policy criticized." *Eugene Register Guard*. December 25, 1992.
"Zoo's oldest Asian elephant to be euthanized." KATU News. August 21, 2006.

About the Author

Melissa Crandall is a writer and essayist whose work has appeared in several publications including *Allegory Magazine*, *Wild Musette*, ASPCA's *Animal Watch Magazine*, and the *Journal of the Elephant Managers Association*. Her short stories "The Cellar" and "Thicker Than Water" were nominated for a Pushcart Prize. Crandall is a member of the Authors Guild and the Elephant Managers Association, and currently resides in Connecticut.

Ooligan Press

Ooligan Press is a student-run publishing house rooted in the rich literary culture of the Pacific Northwest. Founded in 2001 as part of Portland State University's Department of English, Ooligan is dedicated to the art and craft of publishing. Students pursuing master's degrees in book publishing staff the press in an apprenticeship program under the guidance of a core faculty of publishing professionals.

Project Managers
Monique Vieu
Julie Collins

Acquisitions
Ari Mathae
Taylor Thompson
Des Hewson
Kimberley Scofield

Editing
Madison Schultz
Melinda Crouchley

Design
Jenny Kimura
Denise Morales Soto

Digital
Kate Barnes
Megan Crayne Beall

Marketing
Sydney Kiest
Sydnee Chelsey

Social Media
Sadie Verville
Faith Muñoz

Book Production
Stephanie Anderson
Jasmine Gower
Esa Grigsby
Brennah Hale
Zoe LaHaie
Elise Hitchings
Megan Huddleston
Sean Paul Lavine
Hope Levy
Kyle Mader
Laura Mills
Vivian Nguyen
Julianne Pearson
Victoria Raible
Morgan Ramsey
Ruth Robertson
Olivia Rollins
Arielle Roper
Kaitlyn Shehee
Sirisha Vegulla
Emma Wolf
Michelle Zhang